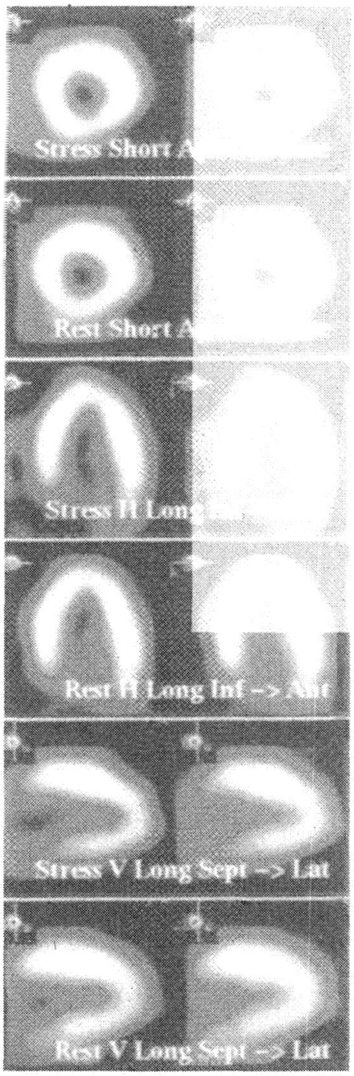

Nuclear Cardiac Imaging

Terminology and Technical Aspects

Elpida S. Crawford, MS, CNMT
Director, Nuclear Medicine Technology Program
School of Medicine and Biomedical Sciences
University at Buffalo
State University of New York
Buffalo, New York

Syed Sajid Husain, MD, MS, MAS
Department of Nuclear Medicine
Center for Positron Emission Tomography
Veterans Affairs Western New York Healthcare System
Director, Nuclear Cardiology
Buffalo Heart Group
Buffalo, New York

snmts
society of nuclear medicine
technologist section

Society of Nuclear Medicine
1850 Samuel Morse Drive
Reston, Virginia 20190-5316

Library of Congress Cataloging-in-Publication Data

Crawford, Elpida S.
Nuclear cardiac imaging: terminology and technical aspects/
Elpida S. Crawford & Syed Sajid Husain.
p. cm.
Includes bibliographical references.
ISBN 0-932004-74-1
1. Heart—Radionuclide imaging. I. Husain, Syed Sajid. II. Title

RC683.5.R33 C73 2002
616.1 297575—dc21

2001049816

Dedication

For all their patience and loving support, I would like to dedicate this book to my husband, Brian, and our children, Stephen and Christine.

—*Elpida S. Crawford*

To my wife, Tahira, for her relentless support and endurance throughout my medical career.

—*Syed Sajid Husain*

Table Of Contents

Foreword

Those of you who have chosen to enter the profession of nuclear cardiology have entered a field that requires native talents and acquired skills that few other specialties demand. Skill in interacting with patients; knowledge of the basic physical, chemical, and biological principles; mastery of complex imaging technology; and safe and accurate handling of radioactive tracers require your best efforts, dedication, and skills.

The story of nuclear cardiology goes back many years. Few realize that the first diagnostic study in nuclear medicine was in cardiology. Herman Blumgart used solutions of radon gas to measure what he called the "velocity of the circulation." Since that time, radioactive tracers and external radiation detection instruments, looking at the heart from outside the living human body, have been developed to measure the pumping function of the heart, its supply of energy, and the control of cardiac function by the autonomic nervous system. Regional myocardial blood flow, metabolism, and innervation can be measured "noninvasively" in normal persons and patients with actual or suspected diseases of the heart.

In 1926, Blumgart was working as a student assistant in the laboratory of Walter Cannon, chair of the physiology department at Harvard Medical School. Blumgart had learned about the treatment of cancer with radium during his fellowship in cardiology in England. He knew that radon gas was released as a daughter product. When he returned to Harvard, he obtained radium discarded after use in radiation therapy and dissolved the escaping radon gas into saline that could be injected into a subject's arm vein. Using a Geiger tube radiation detector, he measured the time it took for the radioactivity to reach the heart in 62 normal volunteers. He subsequently studied patients with thyrotoxicosis, anemia, polycythemia vera, carcinoma, and heart disease. He examined the effects of physiologic and pharmacologic stresses and, with his colleagues, published 22 articles, chiefly in the *Journal of Clinical Investigation.* Marshall Brucer, a pioneer in the field of nuclear medicine, considered the article written by Blumberg with Soma Weiss in 1927 (*J Clin Invest.* 1927;4:389) to be the "first article ever published in the field of nuclear medicine." This article was published the year that Charles Lindbergh made the first flight over the Atlantic Ocean—and the year that I was born.

Subsequent events along the road to nuclear cardiology include Georg de Hevesy's use of red blood cells labeled with ^{32}P to measure red blood cell volume. A Nobel laureate, Hevesy invented the "tracer principle," the most fundamental in nuclear medicine. In 1936, Paul Hahn used ^{59}Fe to measure total body hematocrit. Research during World War II on traumatic shock stimulated considerable research in cardiology and led to the introduction of ^{35}S as a tracer of plasma proteins. Fine and Seligman invented the radioiodine method for measuring plasma volume. By 1943, radioiodinated bovine albumin, the "first true radiopharmaceutical" according to Brucer, had been used in humans. Storaasli replaced bovine albumin with human serum albumin, among the earliest commercially available "radiopharmaceuticals," as radioactive tracers came to be known.

Measuring plasma and red cell volume constituted an addition to two technological advances made many years before: the invention of the stethoscope and the electrocardiogram. When I was a medical student, "heart stations," based on examining the electrical activity of the heart, were becoming commonplace throughout medicine.

The next major breakthrough was catheterization of the human heart by Werner Forssmann in Eberswald, Germany, in 1929. In 1941, Cournand and Richards used cardiac catheterization to measure cardiac output. In 1956, Forssmann, Cournand, and Richards shared the Nobel prize. In his acceptance speech, Cournand stated: "The cardiac catheter was the key in the lock."

Another little known historical fact is that Jeff Holter, the inventor of the Holter monitor to detect cardiac arrhythmias in ambulatory patients, was the person who had the idea to found the Society of Nuclear Medicine in 1954.

In 1976, Gruentzig developed coronary angioplasty for treatment of coronary artery disease. By 2001, 2 million angioplasties had been performed worldwide, with an estimated increase of 8% per year.

As has been the case for nearly all technological advances in cardiology, a therapeutic innovation has been the stimulus for better diagnostic tests to help detect treatable diseases and to plan and monitor treatment. The birth of cardiac surgery shortly after World War II was a major stimulus to the invention of the two most important tools in nuclear medicine, the invention of the "scintillation camera" by Hal Anger in 1958 and the development of ^{99m}Tc, which would become the "workhorse" of the field for decades after it was commercially developed in 1960. The Anger camera was able to image simultaneously the distribution of radiotracers in different parts of the heart. This is in contrast to imaging the distribution serially over time, as was necessary with the rectilinear scanner, invented by Benedict Cassen in 1951 and subsequently used in cardiology only to detect percicardial effusion and diagnose pulmonary embolism. ^{99m}Tc emitted 140 keV photons that were ideal for safely providing large numbers of photons that could be measured with external imaging. Among the first uses was the detection of intracardiac shunts, soon to be followed by measurements of regional myocardial blood flow and regional ventricular function. ^{99m}Tc also had the important advantage of being available from a ^{99}Mo generator, which greatly simplified its availability and commercial development. In 1964, ^{99m}Tc tracers stole the show at international meetings. By 1965, ^{99m}Tc and the Anger camera started the path of logarithmic growth of nuclear medicine. Nuclear cardiology led the way.

In 1979, Berger examined the "first pass" of ^{99m}Tc tracers through the heart to evaluate both right and left ventricular function in patients with coronary heart disease, chronic obstructive lung disease, and congenital heart disease. Global ventricular function, as well as regional contraction of the ventricles, could be examined, together with stroke volumes, ejection fractions, and filling and emptying rates of the ventricles. The invention of single photon emission computed tomography (SPECT) greatly improved the results with planar imaging alone. In 1977, ^{43}K was introduced to measure regional myocardial blood flow, to be replaced eventually by ^{201}Tl. Exercise stress was developed for increasing the sensitivity of coronary heart disease detection.

Although the most important clinical use of positron emission tomography (PET) to date has been in oncology, PET offers many important advantages over SPECT in nuclear cardiology, including better spatial resolution and quantification. Today there are a number of radiopharmaceuticals useful in the study of myocardial perfusion, glucose and fatty acid metabolism, and autonomic innervation of the heart. ^{15}O-water is the most widely used worldwide in nuclear cardiology but requires a cyclotron at the site of the examinations. ^{15}N-ammonia is also useful but has not been commercially available. ^{82}Rb, available from a generator, is likely to be more widely used in the future for examining myocardial perfusion. Regional metabolic imaging with ^{18}F-deoxyglucose is useful in detecting myocardial viability in patients being considered for surgical intervention. Heart failure and better care of patients with arrhythmias are likely to become major foci of nuclear cardiology in the near future. Newer techniques include the detection of plaques on coronary arteries.

The role of the technologist has never been greater in the translation of these exciting, continuing developments in nuclear cardiology. Those persons responsible for the performance of these complex, highly demanding studies can take great pride in their mastery of a field in which excellence in science and patient care is essential.

Henry N. Wagner, Jr., MD
Director, Division of Radiation Health Sciences
Professor Emeritus, Radiology and Radiological Sciences
Professor, Environmental Health Sciences
Johns Hopkins University
Baltimore, Maryland
April 2003

Acknowledgments

A number of colleagues and friends in the Buffalo, NY, area contributed images for this book. Some of these people have mentored me, and some were my students. They include:

Joseph Oliverio, CNMT	Paula Horwitz, CNMT
Jennifer Buckingham, CNMT	Raya Sirtokin, CNMT
Michael Merhige, MD	Fazalur Rehman, MD
George Baeumler, MD	Michelle Roesler, CNMT
Kwang Joo, MD	Rhonda Blatner, CNMT
Paul Galantowicz, CNMT	Kara Niedel, CNMT
Hani Abdel-Nabi, MD, PhD	Kimberly Dix, CNMT
Jayakumari Gona, MD	David Paulin, CNMT
Mark Javier, CNMT	Kelly Molik, CNMT

Other people who assisted us include: John Nyquist, MS, CMI, with medical illustrations, Mary Ellen Kawalec-Suhr for collecting electrocardiograph strips, and Richard Kayton and Frederick Covelli for computer support. Karen Rath, Janet Colurra, and Kathy Bartsch assisted with the typing.

We wish to express our sincere thanks to all of these people.

Elpida Crawford, MS, CNMT
Syed S. Husain, MD, MS, MAS
Buffalo, New York
2003

Chapter 1

Cardiac Anatomy Review

The heart is situated between the lungs in the middle mediastinum. The heart is shaped like a blunt cone about the size of a closed fist. About two-thirds of its mass lies to the left of the body's midline (Fig. 1.1). Its pointed end, the apex, is formed by the tip of the left ventricle. The apex projects inferiorly, anteriorly, and to the left. It lies superior to the central tendon of the diaphragm. Anteriorly, the apex is in the fifth intercostal space. The inferior surface of the heart is in contact with the diaphragm.

Cardiac Anatomy and Review

Pericardium

The heart is enclosed and held in place by the pericardium. The outer fibrous pericardium provides a tough protective membrane around the heart and anchors it in the mediastinum. The inner serous pericardium forms a double layer around the heart. The outer parietal layer is directly beneath the fibrous pericardium. The inner visceral layer, also called the epicardium, is attached to the myocardium (muscle) of the heart. Between the parietal and visceral layers of the serous pericardium is a thin film of up to 150 mL of serous pericardium in the pericardial cavity.

Heart Wall

The wall of the heart is divided into three layers: the epicardium (external layer, also called the serous pericardium), the myocardium, and the endocardium (inner layer). The myocardium, which is thick cardiac muscle, makes up the bulk of the heart and is responsible for the contraction of the heart (see chapters 14 and 22).

Heart Chambers

The interior of the heart is divided into four cavities called chambers that receive circulating blood (Fig. 1.2). The chambers on the right side are separated from those on the left by the septum, a muscular partition. The two superior chambers are the right and left atria. The two inferior chambers are the right and left ventricles.

The right atrium receives blood from all parts of the body except the lungs via the superior and inferior vena cava. From the right atrium, blood passes into the right ventricle and then to the lungs via the pulmonary trunk and arteries to release carbon dioxide and take up oxygen.

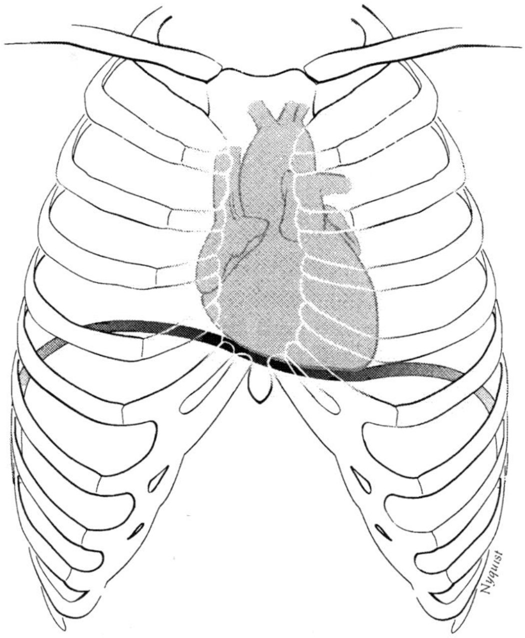

Figure 1.1 Anterior view of the chest, showing the location of the heart.

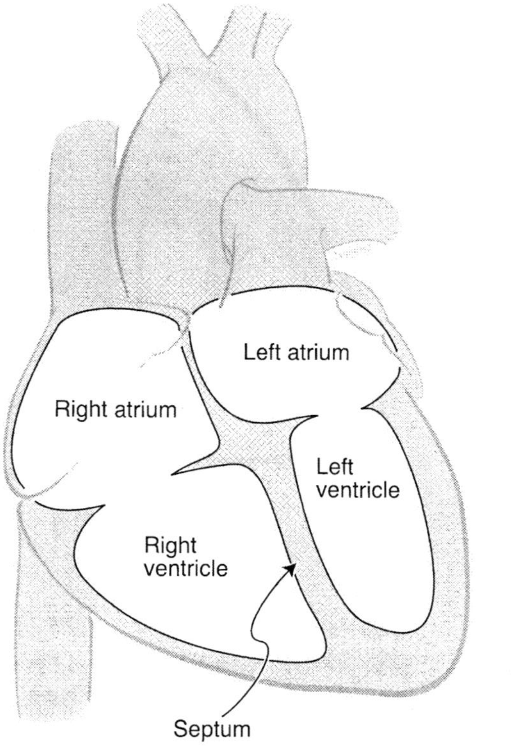

Figure 1.2 Midcoronal view of the heart.

Blood returning from the heart is delivered to the left atrium by the four pulmonary veins. Blood from the left atrium passes into the left ventricle, from which it is pumped into the aorta to be delivered to all parts of the body. Figure 1.3 shows the path of blood through the heart and great vessels. Figures 1.4 and 1.5 are anterior and posterior views of the heart and great vessels.

The myocardium of the ventricles is thicker than that of the atria. The left ventricle wall is much thicker than the right ventricle wall. This is because the left ventricle pumps blood into the aorta for distribution to all parts of the body.

Heart Valves

As each chamber of the heart contracts, it pushes a portion of the blood into a ventricle or out of the heart through an artery. The heart valves keep the blood flowing in the proper direction through the chambers of the heart and prevent backflow of blood. On each side of the heart is an atrioventricular valve. The right atrioventricular valve, also called the tricuspid valve, separates the right atrium from the right ventricle. The left atrioventricular valve, also called the bicuspid or mitral valve, separates the left atrium and left ventricle. The arteries that leave the heart (the pulmonary trunk and aorta) have semilunar valves that prevent blood from flowing back into the heart.

Blood Supply to the Myocardium

The myocardium receives oxygenated blood from its two major coronary arteries, the left main and right main coronary arteries. They branch and encircle the heart to provide blood to all portions of the myocardium. The left main coronary artery branches into the left anterior descending (LAD) and left circumflex arteries. The LAD branch supplies blood to the anterior wall, most of the septum, and the distal portion of the apex. The left circumflex and its branches supply blood to the posterior and lateral walls of the left ventricle. The right coronary artery has multiple branches that supply blood to the inferior wall of the left ventricle and the right side of the heart. Depending on which system (right coronary artery or circumflex) is dominant in a patient there may be an overlap of blood supply to a given territory. Figures 1.4 and 1.5 show the distribution of the coronary arteries to the anterior and posterior surfaces of the heart.

As blood passes through the coronary arteries, it supplies oxygen and nutrients to the myocardium. Most parts of the body receive branches from more than one artery, and these branches may connect to form collateral circulation, which is an alternate route for blood to reach a specific tissue. The myocardium also develops collateral circulation.

Anatomy of the Artery

Arteries transport blood under high pressure and, therefore, possess strong elastic walls. These walls are constructed of three layers. The inner layer is the endothelium. The middle layer, which is the thickest, is composed of elastic fibers and smooth muscle. The outer layer is composed of elastic collagenous fibers. The passageway through which the blood flows is the lumen of the artery.

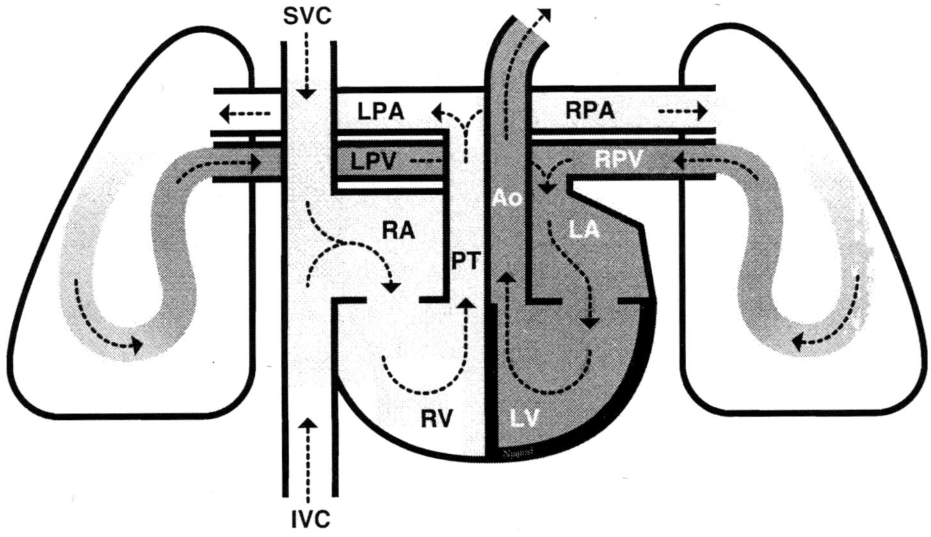

SVC - superior vena cava	**PT** - pulmonary trunk	**RPV** - right pulmonary vein
IVC - inferior vena cava	**LPA** - left pulmonary artery	**LA** - left atrium
RA - right atrium	**RPA** - right pulmonary artery	**LV** - left ventricle
RV - right ventricle	**LPV** - left pulmonary vein	**Ao** - aorta

Figure 1.3 Block diagram showing the pathway of blood through the heart and great vessels.

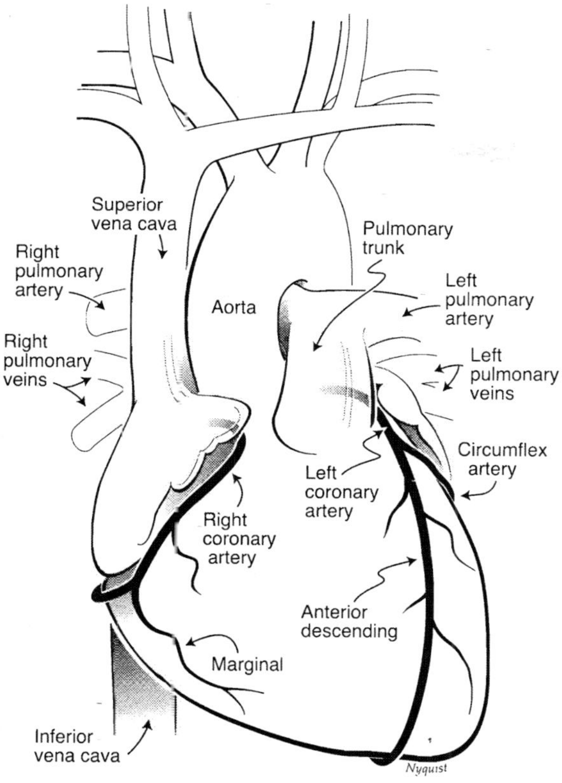

Figure 1.4 Anterior view of the heart, great vessels, and coronary arteries.

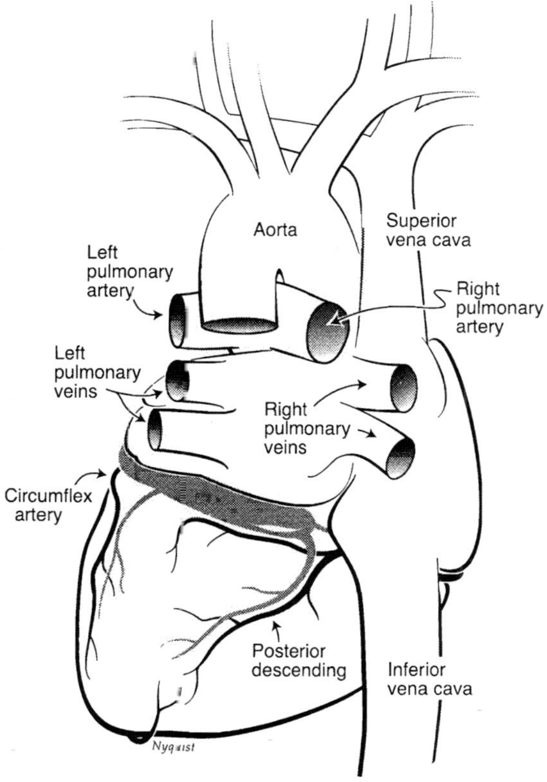

Figure 1.5 Posterior view of the heart, great vessels, and coronary arteries.

Coronary Artery Disease

Cardiovascular disease is the leading cause of death in the United States. It accounts for nearly 1 million deaths each year, half of which are the result of coronary artery disease (CAD). CAD is a condition in which the heart muscle receives an inadequate amount of blood. It is caused by the accumulation of atherosclerotic plaques in coronary arteries. This causes stenosis or narrowing of the lumen of a vessel. In addition, atherosclerotic disease makes the walls of affected vessels less elastic and thus able to dilate to meet demands for increased blood flow, such as during periods of physical exertion or stress. Other causes of coronary occlusion include a thrombus or an embolus in a coronary artery or a coronary artery spasm. This condition of a deficiency of blood resulting from an obstruction or functional constriction of a blood vessel is also referred to as ischemia or ischemic heart disease. Ischemia is the cry of the heart for oxygen and nutrients.

Atherosclerosis

Atherosclerosis is a process in which fatty substances, especially cholesterol and triglycerides (ingested fats), are deposited in the walls of medium- and large-diameter arteries in response to certain stimuli (Fig. 1.6). It is believed that the first event in atherosclerosis is damage to the en-

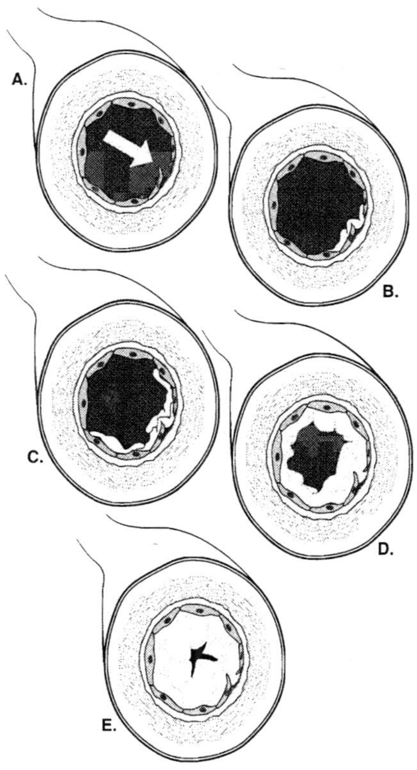

Figure 1.6 Progression of plaque build-up and artery occlusion. (A) The artery lining suffers damage or injury; (B) the injury site provides a rough surface for the beginning of plaque formation; and (C–E) the build-up of plaque progresses.

dothelial lining of the artery. Contributing factors include high blood pressure, carbon monoxide in cigarettes, diabetes mellitus, cytomegalovirus (a common virus), and high cholesterol levels. At the injury site, platelets, phagocytes, and lipids begin to aggregate. The accumulation causes the formation of atherosclerotic plaque. The build-up of plaque is a progressive disease. The more cholesterol in the blood, the faster the process.

The effects of fat in the blood on this process cannot be understated. Cholesterol and other fats are transported in the body by lipoproteins. High-density lipoproteins (HDL) carry cholesterol and other fats away from cells and to the liver for excretion. Low-density lipoproteins (LDL) allow the deposit of fat in tissue and contribute to coronary occlusion. For this reason, cholesterol testing is usually performed as part of a lipid profile test, which also evaluates lipoproteins and triglycerides.

An additional danger of plaque build-up in a vessel is that it may provide a roughened surface that causes blood to release chemicals that may form a thrombus, further obstructing blood flow. Also, a thrombus or piece of a thrombus may dislodge to become an embolus and obstruct blood flow in other vessels.

Risk Factors for Coronary Artery Disease

Factors that increase the risk of developing CAD include high blood cholesterol level, high blood pressure, cigarette smoking, obesity, diabetes, "type A" personality, sedentary lifestyle, family history of CAD, and gender. CAD is more prevalent in males; however, after women reach menopause, no significant gender difference exists in the prevalence of the disease.

Symptoms of Coronary Artery Disease

Because coronary artery narrowing progresses over time, a person may have the disease yet be asymptomatic in the early stages. Symptomatic disease generally manifests itself in one of two ways. The first is called angina pectoris, which means chest pain. The second is a "heart attack," also called myocardial infarction (MI), which means that a portion of the heart muscle dies as a result of inadequate blood supply.

Angina pectoris is a transient pain or discomfort resulting from a temporary lack of adequate blood supply (i.e., oxygen and nutrients) to the heart muscle. The pain is described as a crushing, heavy pressure on the chest and may spread to one (most often the left) or both shoulders, other parts of the upper body, and the epigastric region. Symptoms also may include difficulty in breathing, weakness, dizziness, or perspiration. Symptoms commonly last 2–15 minutes and can be precipitated by such events as exercise, emotional upset, or exertion. As the coronary narrowing increases, the amount of exercise or exertion needed to bring on angina decreases.

In some individuals, especially those with diabetes who suffer from neuropathy, ischemic episodes occur without producing pain. This is known as silent myocardial ischemia and is especially dangerous because the person has no forewarning of an impending heart attack.

Angina pectoris that is new, is worsening in severity, or comes on at rest is called unstable angina. Patients with unstable angina are at high risk for an acute MI.

A heart attack occurs when an area of the heart muscle is deprived of blood (oxygen) for an extended period of time. As a result, the muscle cells in this area die. The infarction can cause altered electrical activity in the heart, which could result in death if not treated promptly.

If the patient survives, infarcted tissue in the heart eventually will be replaced by fibrous tissue. That area of the heart will no longer be able to contract adequately. The overall functional capacity of the heart after an MI will depend on the extent (size) and severity of the MI and whether it was transmural or nontransmural.

Manifestations of Coronary Artery Disease

Before reaching >70%–80% vessel occlusion, CAD may have little or no effect on the heart in a resting stage or during normal nonexertional daily activity. Although the blood flow through the diseased vessels may be adequate for nonexertional activity, it may be quite inadequate when the demands of the heart muscle are increased. Coronary reserve, which is the ability to increase coronary blood flow, decreases in response to increased metabolic demand in diseased coronary arteries. Therefore, in patients with CAD, resting regional myocardial blood flow is usually homogeneous even in the presence of a severely narrowed coronary artery. However, the same degree of narrowing will result in a region of diminished blood flow during stress such as physical exercise (*1*). Collateral circulation may be inadequate when the myocardial demands for oxygen are increased. Collateral vessels depend on the pressure gradient to drive blood flow into the ischemic myocardium. Collateral flow will drop when pressure drops in the artery from which the collateral originates. This reduction in collateral flow is referred to as the coronary steal phenomenon (*2*). Ischemia is manifested when a stress-induced perfusion deficit reperfuses at rest. An understanding of this is key to understanding myocardial perfusion imaging (MPI) and the reasons that MPI is usually performed both with and without stress.

Effect of Coronary Artery Disease on Heart Function

Myocardium subject to chronically reduced blood flow at rest can exhibit a decrease in force of contraction. This may be a protective response of the myocardium to meet the reduced supply of oxygen and substrates. This tissue is referred to as hibernating myocardium. This tissue is viable, and contractile function can be restored with an improvement in blood supply.

Contractile dysfunction also can occur as a result of an acute ischemic episode. Postischemic dysfunction that persists after normal blood flow has been restored is called myocardial stunning. Stunned myocardium eventually returns to normal without medical intervention. The postischemic dysfunction, however, may last from days to weeks. Situations that may cause myocardial stunning include cardiac surgery, coronary angioplasty, acute coronary occlusion followed by thrombolysis, exercise-induced ischemia, and unstable angina (*3,4,5*). Table 1.1 provides a basic comparison of hibernation and stunning.

Treatment Options for Coronary Artery Disease

In addition to the risks of heart attack and/or death, CAD can severely affect the quality of a person's life. Noninvasive treatment options for CAD include drug therapies, low-fat diet, and exercise. Drugs used for CAD include nitroglycerin, beta blockers, cholesterol-lowering drugs, and clot-dissolving agents. Nitroglycerin relaxes smooth muscle, causing blood vessels to dilate. Beta blockers depress cardiac function and decrease cardiac output.

Commonly employed revascularization procedures in patients with CAD include coronary artery bypass graft (CABG) surgery, percutaneous transluminal coronary angioplasty (PTCA), and adjunctive use of intracoronary stents and atherectomy.

CABG is a surgical procedure in which blood vessels from another part of the body are used to bypass a blocked region of a coronary artery to improve the blood supply to an area of ischemic muscle at risk for infarction. Vessels commonly used for revascularization are the saphenous vein and the internal mammary arteries.

PTCA was first performed in 1977. There has been a tremendous growth in the number of coronary angioplasty procedures in the last 2 decades. The number of PTCAs performed in the United States has far exceeded the number of CABG procedures. The transluminal angioplasty is a technique by which the lumen of a stenotic vessel is dilated. As an adjunct to PTCA, innovative technologies such as atherectomy, intracoronary stents, and lasers have played important roles in improving the outcome of coronary revascularization procedures.

The basic mechanism of angioplasty involves positioning a deflated balloon across the stenosis. The balloon catheter within the stenotic segment is then inflated. The inflation of the balloon catheter results in cracking of the intimal plaque and stretching of the media and adventitia, causing expansion of the outer diameter of the stenotic coronary artery. The adjunctive use of atherectomy and stents reduces the incidence of restenosis after balloon angioplasty.

Coronary atherectomy is a procedure in which the obstructive plaque or thrombus is removed by atherectomy devices. In atherectomy, the coronary lumen is enlarged by debulking the culprit lesion. These devices may be ablative lasers, which remove plaque by tissue vaporization, or mechanical high-speed rotation devices for cutting atherosclerotic plaque. The two most common types of devices and procedures used are directional coronary atherectomy and rotational atherectomy (rotablator atherectomy).

Intracoronary stents are the treatment of choice for acute or threatened closure or restenosis of the dilated coronary artery. Stenting is a promising approach to optimize the results of primary angioplasty for acute MI and to treat complications. Currently, stenting is performed in >70% of procedures. There are a variety of different types of stents. The two major types of intracoronary stents are those that are balloon expandable and those that are self expandable. Stent designs include mesh, wire coil, wire mesh, slotted tube, or zigzag. The stent is placed after balloon angioplasty, which creates intimal disruption. After placement, the stent is expanded either by self expansion or balloon expansion, depending upon the type of stent used. Figure 1.7 is a diagram of the placement of a coronary stent.

Clinical indications for PTCA include significant stenosis of one or more major coronary arteries, recurrent angina after MI, or angina with objective evidence of myocardial ischemia that has not responded adequately to medical therapy.

Each treatment option has its advantages, disadvantages, and limitations.

Indications for Myocardial Perfusion Imaging

MPI is a useful, noninvasive method for detecting CAD, assessing the functional significance of coronary stenosis, evaluating prognosis, and assessing therapy. MPI is most often performed in conjunction with stress (either physical or pharmacologic) to induce coronary ischemia (if, in fact, it is present). A radiopharmaceutical that will distribute in the myocardium proportionally to blood flow is injected into the patient at peak stress. Images of the heart are acquired to compare the blood supply at stress with the blood supply at rest. Images taken at rest may require a second injection, depending on the radiopharmaceutical used. These images can identify areas and extent of normal blood flow and reversible (ischemic) or irreversible defects in the heart.

TABLE 1.1

Comparison of Hibernating and Stunned Myocardium

	Hibernating	Stunned
Blood flow	Decreased	Normal
Contractile function	Decreased	Decreased
Necessity of medical intervention to restore function	Yes	No

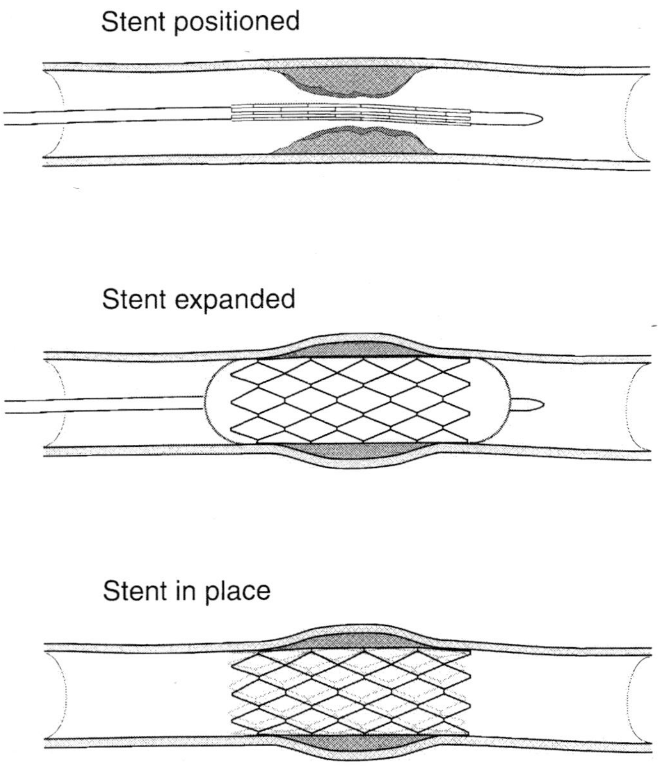

Figure 1.7 Diagram of a coronary stent.

Patients with a normal myocardial perfusion study have excellent prognoses, despite the fact that they may have underlying CAD. A normal stress–rest study predicts a very low (<1%) likelihood of cardiac death or MI for at least the next 12 months. This prediction is independent of gender, age, symptoms, and other risk factors. On the other hand, patients with an abnormal myocardial perfusion study are at increased risk for adverse cardiac events (*6,7*). MPI also is used for risk assessment of patients after an MI.

MPI helps to identify which patients will benefit from medical therapies or invasive revascularization procedures such as CABG and angioplasty. MPI is used to evaluate the effectiveness of these and other medical therapies and lifestyle modifications.

Nuclear cardiac studies also are used for preoperative evaluation of selected patients recommended for noncardiac surgery to assess cardiac risk. In some patients, the presence of CAD may be obvious or known. On the other hand, many patients without cardiac symptoms may have severe CAD that is not obvious because these individuals are functionally limited by severe arthritis or peripheral vascular disease (*8,9*).

Many patients each year visit the emergency room with acute chest pain. A growing number of studies indicates that MPI at rest is very useful in the triage of these patients, leading to better patient management and a reduction in the overall costs for this area of care (see chapter 13).

References

1. Zaret BL, Walkers FJ. Nuclear cardiology—part one. *New Engl J Med.* 1993;329:775–783.
2. Patterson RE, Martin SE, Cedarholm JC. Pathophysiology of coronary artery disease as applied to nuclear imaging. In: Guiberteau MJ, ed. *Nuclear Cardiovascular Imaging.* 1st ed. New York, NY: Churchill Livingstone, Inc.; 1990:9–10.
3. Bolli R. Myocardial stunning in man. *Circulation.* 1992;86:1671–1691.
4. Marban E. Myocardial stunning and hibernation. The physiology behind the colloquialism. *Circulation.* 1991;83:681–688.
5. Mesotten L, Maes A, Hambye A, et al. Nuclear cardiology. Part IV. Viability. *J Nucl Med Technol.* 1999;27:93–102.
6. Chatziioannou S, Moore W, Ford P, et al. Prognostic value of myocardial perfusion imaging in patients with high exercise tolerance. *Circulation.* 1999;99:867–872.
7. Bateman T. Clinical relevance of a normal myocardial perfusion scintigraphic study. *J Nucl Cardiol.* 1997;4:172–173.
8. Eagle K, Rihal C, Mickel MC, et al. Cardiac risk of noncardiac surgery. *Circulation.* 1997;96:1882–1887.
9. Eagle K, Brundage B, Chaitman B, et al. Guidelines for perioperative cardiovascular evaluation for noncardiac surgery. *Circulation.* 1996;93:1278–1317.

Chapter 2

Radiopharmaceuticals

Radiopharmaceuticals currently used in nuclear medicine cardiac perfusion studies are reviewed in this chapter. Included are descriptions of physical properties, biodistribution, preparation, and dose ranges, as well as assessments of advantages and disadvantages.

Thallium-201

Thallium-201 (^{201}Tl), a cyclotron-produced potassium (K+) analog, was first introduced as a myocardial perfusion imaging agent in the early 1970s (*1*). Before the introduction of thallium, the assessment of myocardium perfusion was a cumbersome and somewhat invasive procedure involving the selective intracoronary injection of technetium-99m (^{99m}Tc)-labeled microspheres or macroaggregated serum albumin.

Physical Properties

^{201}Tl decays by electron capture to mercury-201 (^{201}Hg), with a physical half-life of 73.1 hours. ^{201}Tl emits two gamma rays at energies of 135 and 167 keV. However, the principal photons for imaging are the more abundant ^{201}Hg x-rays, which are in the 68–80 keV range.

Biodistribution

As a potassium analog, thallium can substitute for potassium and accumulate rapidly in living cells. Biodistribution is generally proportional to organ blood flow at the time of injection. Myocardial uptake plateaus at high flow levels (four times resting blood flow) (*2*). ^{201}Tl concentration in the blood decays exponentially with a half-time of <30 seconds, decreasing to 2%–3% of the injected dose by 8–10 minutes (*3*). Blood clearance is primarily through the myocardium, kidneys, thyroid, liver, and stomach, with the remainder distributing fairly uniformly through the body. Maximum myocardial uptake of 4%–5% is reached within 5–10 minutes.

A small amount of ^{201}Tl (8%–10%) normally is deposited in the lungs. Increased pulmonary uptake of thallium is seen with increased left atrial pressure and prolonged pulmonary transit time (*4*).

Initial Myocardial Uptake

^{201}Tl is known for its high extraction efficiency by myocardial cells. The extraction of thallium by the myocardium is by the active transport mechanism of the sodium (Na)–K adenosine triphosphatase (ATPase) system. Thallium has a relatively prolonged clearance time from the myocardium when compared with potassium (*3*). The relatively prolonged clearance time facilitates the imaging of myocardium by gamma cameras.

The initial distribution of ^{201}Tl in the myocardium is proportional to regional myocardial blood flow at the time of injection. If ^{201}Tl is administered at the time of myocardial ischemia, more will be delivered to normally perfused tissue than to the ischemic myocardium. Thallium usually does not distribute in infarcted or scarred tissue.

Redistribution

The distribution pattern of ^{201}Tl in the myocardium is not stable but changes as a function of time. The initial uptake of thallium in the myocardium soon after intravenous administration is a function of myocardial perfusion and the extraction of the tracer by the myocardium. Thallium does not remain fixed in the myocardium but continuously exchanges with new thallium from systemic recirculation. This process of thallium exchange or redistribution after transient hypoperfusion brings the thallium into a state of equilibrium within the myocardium. The variables that affect redistribution of thallium in viable myocardium include the washout rate of thallium and the coronary perfusion pressure. Redistribution of thallium begins immediately after the initial uptake. Over a period of time, the intracellular thallium concentration will normalize and all viable myocardial cells will reach a state of equilibrium (*5*).

Reduced ^{201}Tl uptake in a region of the myocardium immediately after exercise that redistributes (or fills in) within 3–4 hours is considered to represent ischemic myocardium. An exercise-induced perfusion deficit that does not return to normal within 3–4 hours is considered an irreversible perfusion deficit. Irreversible perfusion deficits may represent myocardial infarction or hibernating myocardium and require further investigation. Delayed imaging (up to 24 hours) may help to distinguish viable from nonviable myocardium. Other approaches to the resolution of irreversible defects are discussed in chapter 6.

Preparation and Dose Range

^{201}Tl is supplied in the form of thallous choride-201. Storage is at room temperature. The recommended dose per injection ranges from 3–4 mCi, depending on the specific procedure.

Agents Labeled with Technetium-99m

Physical Properties

^{99m}Tc is generator produced. Its parent nuclide is molybdenum-99 (^{99}Mo). ^{99}Mo decays by beta emission; 87% of its decay goes to the metastable-state ^{99m}Tc and the remaining 13% to the ground-state ^{99}Tc. ^{99m}Tc has a half-life of 6 hours and decays to ^{99}Tc by isomeric transition with one gamma ray of 140 keV (*7*).

Technetium-99m–Sestamibi

Description and Biodistribution

^{99m}Tc-sestamibi became commercially available in the United States in 1990. This agent belongs to a class of compounds called isonitriles. Myocardial uptake of sestamibi after intravenous injection is proportional to myocardial blood flow and viability of the tissue, similar to ^{201}Tl. The myocardial extraction fraction of sestamibi is somewhat lower than that of ^{201}Tl (*2*). Myocardial uptake of ^{99m}Tc-sestamibi is approximately 1.2%–1.5% of the injected dose.

The intracellular extraction of sestamibi is less dependent on active transport mechanisms than is that of ^{201}Tl. Cellular uptake and retention of sestamibi are dependent on passive diffusion of sestamibi across the plasma and mitrochondrial membranes. After extraction, sestamibi remains relatively fixed in myocardial cells, with minimal delayed redistribution (*8*).

Blood clearance is fairly rapid. Pulmonary activity is minimal even immediately after injection. The major pathway for clearance of ^{99m}Tc-sestamibi is the hepatobiliary system. Activity from the gallbladder appears in the intestines within 1 hour of injection. Postinjection imaging time is selected to maximize the count ratio between the heart and surrounding organs (i.e., when liver and large bowel activity is least).

Preparation and Dose Range

Sestamibi is supplied in a kit, which is stored at room temperature before and after reconstitution. Labeling is accomplished by adding the required amount of technetium pertechnetate ($^{99m}TcO_4$) and then heating the kit vial in a boiling water bath for 10 minutes (see kit instructions). Radiochemical purity is determined by chromographic technique.

The recommended dose ranges from 10–30 mCi.

Technetium-99m–Tetrofosmin

Description and Biodistribution

^{99m}Tc-tetrofosmin, which became commercially available in the United States in 1996, is a diphosphine compound. After intravenous administration, this compound is rapidly cleared from the blood and is taken up by the heart, skeletal muscle, liver, spleen, and kidneys proportionally to blood flow. Approximately 1.2% of the injected dose goes into the myocardium. Heart uptake, retention, and blood clearance of tetrofosmin are similar to those of sestamibi. Once tetrofosmin is taken up by the myocardium, there is little, if any, redistribution (*9,10,11*). Liver clearance of tetrofosmin is faster than that of sestamibi (*10,12*).

Preparation and Dose

Tetrofosmin is supplied in a kit that is refrigerated (2°–8° C) before reconstitution. After adding the required amount of $^{99m}TcO_4$, the kit is incubated at room temperature for 15 minutes. The reconstituted kit should be stored at 2°–25° C. Radiochemical purity is evaluated by a chromatographic procedure. The kit should be protected from light. The shelf life of the reconstituted kit is 12 hours.

The recommended dose ranges from 5–24 mCi.

Comparison of Thallium, Sestamibi, and Tetrofosmin

Thallium-201

The ability of thallium to reperfuse and fill in ischemic regions of the myocardium is significant in the diagnosis and evaluation of coronary artery disease (CAD). Because redistribution of thallium begins immediately, stress imaging must begin as soon as possible (within 5–10 minutes) after injection. The redistribution of thallium is advantageous in terms of having one injection for both stress and rest imaging for the detection of CAD. Thallium is a preferred agent for the assessment of myocardial viability by virtue of its active Na-K ATPase transport mechanism and redistribution.

Increased lung uptake of thallium after exercise is a good marker for left ventricular dysfunction resulting from CAD. Multiple perfusion defects and abnormal thallium lung uptake are associated with high risk for cardiac events (*13,14*).

Thallium's physical properties are its greatest limitation. The low photon energy (68–80 keV) limits image resolution. The relatively long half-life and unfavorable dosimetry limit the dose administered. The lower count rate from the limited dose and the redistribution property are not optimal for a gated single-proton emission computed tomography (SPECT) study.

Sestamibi and Tetrofosmin

These agents remain relatively fixed after myocardial uptake, accommodating more flexibility in image acquisition (30–120 minutes) after injection. ^{99m}Tc has a more suitable photon energy (140 keV) for image resolution and a shorter half-life. The doses of these agents are larger, which results in increased photon flux, improving resolution and facilitating gated SPECT. The larger dose also accommodates the acquisition of a first-pass study from which a right and/or left ventricle ejection fraction can be obtained (see chapters 15 and 20).

TABLE 2.1

Advantages and Disadvantages of Currently Available SPECT Perfusion Agents

	Advantages	Disadvantages
Thallium	Reperfusion into ischemic but viable tissue over time. Increased lung uptake after exercise is a marker of coronary artery disease.	Stress imaging must begin soon (5–10 min) after injection. Low photon energy (68–80 keV) limits resolution. Relatively long half-life (73 h) limits the dose administered. Redistribution and limited dose (3–4 mCi) minimize obtaining an optimal gated SPECT study.
Sestamibi	Remains relatively fixed in the myocardium, accommodating flexibility in image acquisition time after injection (30–120 min). Larger dose administered resulting in increased count rate and improved resolution. More suitable photon energy for image resolution and short half-life. Well suited for gated SPECT acquisition. Dose is compatible with first-pass acquisition.	Two doses are needed for a stress and rest study. Liver uptake can cause image artifacts.
Tetrofosmin	Faster liver clearance than sestamibi, accommodating an earlier imaging time after injection. All the advantages listed above for sestamibi.	Two doses are needed for a stress and rest study. Liver uptake can cause image artifacts.

Two doses are always required for each of these agents—one for the stress study and one for the rest study. Subdiaphragmatic activity resulting in liver, biliary, and bowel uptake can cause image artifacts. Tetrofosmin has a faster liver clearance than sestamibi, accommodating an earlier imaging time after injection.

Table 2.1 summarizes the advantages and disadvantages of these agents.

Technetium-99m–Teboroxime

This agent was approved by the U.S. Food and Drug Administration around the same time as sestamibi.

Description and Biodistribution

Teboroxime is chemically very different from thallium and the other technetium agents. Teboroxime is a neutral lipophilic boronic acid (BATO) compound. Myocardial uptake of this agent parallels flow in a linear fashion, without plateauing at high coronary flow rates as do the other agents. The uptake in the myocardium is also higher than the mean thallium extraction (*2*). However, myocardial clearance is very rapid. There is no redistribution of this agent in the myocardium. The liver is the major route for metabolism. The large bowel is the major organ for excretion of teboroxime (*15,16*).

Advantages/Limitations of Teboroxime

As with the other ^{99m}Tc agents, two separate doses are required to complete the stress and rest studies. An optimal imaging protocol requires prompt onset of imaging after tracer injection and a short image acquisition time (5–10 minutes). SPECT imaging is best achieved with a multihead camera. Pharmacologic stress may be preferred over exercise to facilitate immediate postinjection image acquisition.

Liver uptake may interfere with inferior wall visualization. Because of the rapidly changing tracer concentration, gated SPECT is not an option. The advantage, if any, of this agent is that sequential stress and rest myocardial imaging studies can be performed in less than 1.5 hours because of its relatively rapid uptake and washout from the myocardium (*2*).

References

1. Lebowitz E, Greene MW, Fairchild R, et al. Thallium-201 for medical use. *J Nucl Med.* 1975;16:515–155.
2. Berman D, Hosen K, Maddahi J. The new ^{99m}Tc myocardial perfusion imaging agents: ^{99m}Tc-sestamibi and ^{99m}Tc-teboroxime. *Circulation.* 1991;84(suppl 1):7–21.
3. Strauss H, Pitt B. Thallium-201 as a myocardial imaging agent. *Semin Nucl Med.* 1977;7:49–58.

4. Pohost G, Alpert N, Igwall J, et al. Thallium redistribution mechanisms and clinical utility. *Semin Nucl Med.* 1980;10:70–93.

5. Grunwald A, Watson D, Holzgrefe H, et al. Myocardial thallium-201 kinetics in normal and ischemic myocardium. *Circulation.* 1981;64:610–617.

6. Dilsizian V, Rocco T, Freedman N, et al. Enhanced detection of ischemic but viable myocardium by the reinjection of thallium after stress–redistribution imaging. *N Engl J Med.* 1991;122:613–619.

7. Saha G. *Fundamentals of Nuclear Pharmacy.* 3rd ed. New York, NY: Springer-Verlag; 1992:68.

8. Beller G, Glover D, Edwards N, et al. ^{99m}Tc-sestamibi uptake and retention during myocardial ischemia and reperfusion. *Circulation.* 1993;87:2033–2042.

9. Jain D, Walker F, Mattera J, et al. Biokinetics of technetium-99m tetrofosmin myocardial perfusion imaging agent. *J Nucl Med.* 1993;34:1254–1259.

10. Higley B, Smith F, Smith T, et al. Technetium-99m-1, 2-bis[bis(2-ethoxyethly) phosphine] ethane: human biodistribution, dosimetry and safety of a new myocardial perfusion imaging agent. *J Nucl Med.* 1993;34:30–38.

11. Zaret B, Rigo P, Walker F, et al. Myocardial perfusion imaging with ^{99m}Tc tetrofosmin. *Circulation.* 1995;91:313–319.

12. Munch G, Neverve J, Matsunari I, et al. Myocardial technetium-99m-tetrofosmin and technetium-99m-sestamibi kinetics in normal subjects and patients with coronary artery disease. *J Nucl Med.* 1997;38:428–432.

13. Brown K. The role of stress redistribution thallium-201 myocardial perfusion imaging in evaluating coronary artery disease and perioperative risk. *J Nucl Med.* 1994;35:703–706.

14. Beller G. Myocardial perfusion imaging with thallium-201. *J Nucl Med.* 1994;35:674–680.

15. Johnson L. Clinical experience with technetium-99m-teboroxime. *Semin Nucl Med.* 1991;21:182–189.

16. Johnson LL. Myocardial perfusion imaging with technetium-99m-teboroxime. *J Nucl Med.* 1994;35:689–692.

Chapter 3

Exercise Stress Testing

Exercise testing is an established noninvasive tool in the diagnosis of coronary artery disease (CAD). Exercise-induced typical anginal chest pain is a good indicator for the presence of CAD. The presence of diagnostic ST depression in conjunction with treadmill-induced anginal symptoms is highly predictive of coronary disease. Exercise-induced symptoms are predictive of CAD, but not all patients have typical anginal symptoms and/or ST diagnostic segment changes during exercise (*1*). As noted in chapter 1, comparison of stress myocardial perfusion imaging (MPI) with a resting imaging study provides a means of identifying CAD that might otherwise be missed by an exercise test alone. MPI provides valuable information about the extent and severity of a hemodynamically significant CAD. Compared with the exercise tolerance test alone, it significantly improves the specificity of the detection of CAD.

Treadmill Exercise: Description and Guidelines

The treadmill is the most common device used for exercise testing (*2*). A treadmill is a motorized exercise device with a moving belt mechanism that permits walking or running in a stationary location. The exercise level can be adjusted by changing the speed and/or incline of the belt platform.

Although many different exercise procedures may be used, basic equipment, personnel, and criteria are necessary to conduct a valid diagnostic test and ensure the comfort and safety of the patient. The American Heart Association (AHA) provides guidelines for clinical exercise testing laboratories (*3*). These guidelines describe the testing room environment, which should be large enough to accommodate all the equipment necessary and provide walking areas and adequate access to the patient in emergency situations. The room should be well ventilated. Windows and/or wall posters can reduce boredom and anxiety.

The AHA guidelines also describe specifications for a suitable treadmill and other necessary equipment. The treadmill should have padded front and side rails for patient safety and stability, if needed. An emergency stop button should be visible and accessible to the staff and patient. Other essentials include a suitable electrocardiographic (ECG) continuous monitoring system, blood pressure monitoring equipment, and emergency equipment and drugs. The ECG recording device should have both a paper-strip chart recorder and an oscilloscope for continuous visual monitoring. Several sizes of blood pressure cuffs should be available. A cardiac defibrillator and all necessary drugs should be on hand for the treatment of any unexpected cardiac emergency. The room should have a wall-mounted clock with a second hand or digital counter.

A qualified physician should supervise exercise stress tests for cardiac patients, and all procedures should be performed in accordance with AHA and American College of Cardiology guidelines. All staff involved in the performance of exercise stress testing must be trained in basic life support. Advanced life support is required for the supervising physician and strongly recommended for other staff (*3,4*).

Many different exercise protocols exist. By convention, exercise capacity is reported in metabolic equivalents (METS) and not in treadmill time, so that exercise capacity can be uniformly compared between protocols. A MET unit is the energy expenditure at rest equivalent to an oxygen uptake of approximately 3.5 mL O_2/min/kg body weight. Two or three times the oxygen consumption at rest equals 2 or 3 METS, respectively. Commonly used exercise protocols have been developed so that each workload approximates a multiple of the oxygen consumption at rest. Patients limited by symptoms usually have a workload capacity of about 6 METS or less. Asymptomatic cardiac patients may have performance capacities ranging from 7–10 METS or more. Regularly active healthy men may have capacities of 12–15 METS, and athletes may range from 16–20 METS (*2,5*).

The Bruce protocol is most often used in North America. This protocol consists of seven 3-minute stages. At each stage, the patient walks on the treadmill, which has been set to a specific combination of speed and incline (Table 3.1).

Patient Selection for an Exercise Study

The exercise stress test is generally preferred to other types of stress tests when a patient is capable of performing adequate exercise. Candidates should be able to participate in the exercise protocol long enough to achieve 85% of their maximum predicted heart rate (HR) in order to have an optimal exercise tolerance test. The maximum predicted HR is 220 minus the patient's age (Table 3.2). If a patient cannot achieve at least 85% of his or her maximum pre-

TABLE 3.1

Bruce Protocol

Stage	Duration (min)	Speed (mph)	Grade (%)	Metabolic equivalents
I	3	1.7	10	5
II	3	2.5	12	7
III	3	3.4	14	10
IV	3	4.2	16	13
V	3	5.0	18	16
VI	3	5.5	20	19
VII	3	6.0	22	22

dicted HR, there is a risk that CAD will go undetected because the exercise is inadequate to achieve the maximal coronary blood flow and, thereby, create an imbalance between normal and abnormal perfused myocardial segments. Exercise should be symptom limited, with patients achieving >85% of their predicted target HR for optimal diagnostic accuracy (*6*). The symptoms associated with stress-induced ischemia include chest pain, extreme fatigue, arrhythmias, falling systolic blood pressure, and ST segment changes. For patients who have recently had a myocardial infarction (MI), peak exercise levels are usually limited to an HR of 120–130/min.

Patient Preparation for an Exercise Study

Patients should be instructed to dress comfortably and wear walking shoes. They should bring a list of all medications they are taking on the day of the test. Whenever possible, beta blocker medications should be stopped 24–48 hours before a stress study. A patient on beta blocker medication will not achieve 85% of the predicted maximum HR. Patients should be fasting for at least 3–4 hours before exercise. Diabetic patients should be carefully evaluated and instructed appropriately about fasting.

TABLE 3.2

Examples of Estimated Heart Rate for Stress Testing

Patient age (y)	Maximum predicted heart rate* (bpm)	85% of maximum predicted heart rate (bpm)
35	185	157
50	170	145
62	158	134
78	142	121

* Calculated as 220 minus patient age in years.

An intravenous line is secured in the patient's arm before exercise. The radiopharmaceutical will be administered through the line at the time of peak stress. If the dose is injected directly into a vein and not through an established intravenous line, there is a much greater chance that the dose will infiltrate and compromise the validity of the study.

A good procedure explanation is necessary so that the patient will understand what is expected of them. The patient will be expected to exercise until ischemic symptoms occur. The patient needs to communicate honestly with the staff about symptoms and tolerance for continued exercise. The radiopharmaceutical should be administered near the peak of maximum symptoms. After administration of the dose, exercise should continue for at least 1 additional minute to allow tracer uptake in the myocardium.

The patient is prepared for 12-lead ECG monitoring during the study. Proper skin prep is essential to secure a noise-free signal. Hair in the general region of electrode placement should be shaved. Also, the skin should be cleansed with an alcohol-saturated gauze pad or other abrasive to remove superficial dead skin and oils.

An informed consent for exercise stress testing must be obtained before starting the study. A sample consent form is shown in Figure 3.1.

Exercise Perfusion Study Procedure

Resting ECG, HR, and blood pressure should be measured in both supine and standing positions before testing to determine the presence of any abnormalities. For stable patients, the stress test is typically performed with the Bruce protocol. A modified Bruce or other less strenuous protocol is used for unstable patients or those recovering from a recent MI. The patient's blood pressure is taken every minute during the test. The physician monitors the patient's symptoms, HR, and ECG abnormalities during the test.

The radiopharmaceutical is injected through the intravenous line at the attainable level of peak exercise. The line should be flushed immediately with saline. As stated previously, the patient is encouraged to continue exercis-

Figure 3.1 Sample consent for exercise tolerance/nuclear test.

Consent for Exercise Tolerance/Nuclear Test

To facilitate the diagnosis and appropriate management of my condition, I hereby voluntarily consent to engage in an exercise test to help evaluate the functional state of my heart and circulation. Before undergoing the test, I shall be interviewed by a physician in order to determine if I have any condition that would indicate that I should not engage in this test.

I understand that I shall walk on a motor-driven treadmill with progressive increments in speed and incline. During the exercise, my electrocardiogram will be continuously monitored and my blood pressure periodically recorded. The exercise will be discontinued when I attain a predetermined end point, develop any abnormal change that the physician feels is significant, or have distressing symptoms such as chest discomfort, shortness of breath, or fatigue.

I understand that every effort will be made to conduct the exercise test in such a way as to minimize discomfort or risk, although such tests are occasionally associated with certain changes during or just after the procedure. These include episodes of transient lightheadedness, fainting, chest discomfort, changes in the heart rhythm, and, in exceptionally rare instances, heart attacks. I understand that the laboratory is properly equipped with emergency equipment and that trained personnel are available to deal with any unusual situations that may arise.

I have read the foregoing and understand it. Any questions that may have occurred to me have been answered to my satisfaction.

____
Patient's signature

____
Witness

____
Date

There is no chance that I am pregnant or breastfeeding.

____
Patient's signature

____
Date

Figure 3.2 Sample emergency notification procedure.

Accidental Spills of Radioactive Materials
Nuclear Medicine Cardiac Stress Lab

Two radiotracers are commonly used in the nuclear medicine cardiac stress lab to evaluate cardiac function. These radiotracers are technetium (^{99m}Tc)-labeled myocardial perfusion imaging (MPI) agents and thallium (^{201}Tl). Spills of radioactive materials will be handled according to the procedures outlined in the Radiation Safety Protection Manual (RPPM).

Notice: If a patient is in motion on the exercise treadmill when a spill occurs, the safety of the patient will not be compromised. Once the treadmill has stopped, all efforts can then be taken to contain the spill. Major spills will result in the closing of the stress lab and canceling patients under the direction of the radiation safety officer (RSO). In most cases, a minor spill has a greater potential of occurring. Therefore, containing the spill, determining contamination, and starting decontamination procedures as soon as possible will be most practical and will result in the lowest exposure to all involved.

(A) Minor Spills

Cover the spill with absorbent paper to prevent further contamination. Monitor all equipment and personnel involved. If contamination is detected, start decontamination procedures as outlined in the RPPM. Place adequate shielding over areas with exposure readings >0.2 mR/h for all nonremovable contamination areas. Allow the shielded areas to decay 10 half-lives as outlined below. Document the incident with survey results on the Survey Report Form. The RSO must be notified immediately when contamination of personnel is confirmed.

Decay Table

Isotope	Half-life	10 half-life decay time
^{99m}Tc	6 hours	60 hours (2.5 days)
^{201}Tl	3.4 days	34 days

(B) Major Spills

Close off area, lock all doors, and post appropriate signs warning against entry. Allow the area to decay 10 half-lives. Monitor all personnel involved and start decontamination procedure if necessary, monitor adjacent areas to determine exposure rates, and notify the RSO immediately. Document the incident on the Accident Form and the Lab Survey Report Form.

RADIATION SAFETY OFFICER

Name: ______________________ Office phone: ______________________

Home phone: ______________________

Alternate names and telephone numbers designated by the RSO:

CHIEF TECHNOLOGIST

Name: ______________________ Office phone: ______________________

Home phone: ______________________

ing for an additional minute after the injection of the radiotracer. If exercise must be terminated prematurely, the reasons (shortness of breath, hypotension, arrhythmias, chest pain, and/or other symptoms) should be noted (*7*).

Radiation Safety in the Stress Lab

The radioactive materials license application of a facility that performs MPI must identify the location of the stress lab in which radioactive materials will be used and/or stored. If the location of the stress lab is changed, the license should be amended to reflect the change. A Geiger-Müller (GM) survey meter should be kept in this lab. The room should have a sink and a tile floor (not carpeting) to facilitate the clean-up of radioactive spills.

If radioactive waste (contaminated needles, tubing, and other items) is stored in the stress lab, the door of the lab must be locked at all times when trained personnel are not in the room, or the radioactive waste container should be locked. The stress lab door should be posted with the appropriate radioactive materials sign. An emergency notification sign listing spill procedures and the radiation safety officer's (RSO's) telephone number should be posted inside the lab. A sample emergency notification procedure is provided in Figure 3.2.

If the stress lab is a shared lab used by personnel not trained in the handling of radioactive materials, it is recommended that all waste material be removed from the room and stored in the nuclear medicine department hot lab. If not, the waste storage area should be locked, as noted previously.

All personnel (physicians, nurses, technologists, and others) involved in the performance of MPI stress studies should be issued whole-body badges, which should be worn while working in the stress lab. Anyone who handles radioactive materials also should be issued a ring dosimeter. All personnel should be trained in radioactive materials handling commensurate with their level of participation in the study.

Because the radiopharmaceutical is administered to the patient at peak exercise, extra caution should be taken to avoid leakage of the dose onto the treadmill, floor, or other items in the room. One method used to avoid contamination during injection is to place a chux below the injection area. Should a spill occur, it should be appropriately contained and cleaned up. The safety of the patient should be attended to first, and the patient should be removed from the treadmill as soon as possible after the completion of exercise.

The RSO should be notified according to the posted emergency sign. The decontamination effort should continue until the level of removable contamination is <200 dpm/100 cm^2. The treadmill does not need to be closed down if the fixed contamination as measured by a survey meter is <2 mR/h at a distance of 30 cm.

Treadmills used for both radioactive and nonradioactive studies should be surveyed with a GM survey meter after *each* radioactive study. If an unknown spill has occurred, the treadmill should be decontaminated before the next patient use. (Most standard procedures for other nuclear medicine studies require only end-of-the-day monitoring.)

References

1. Pina IL, Madonna DW, Sinnamon EA. Exercise test interpretation. *Cardiol Clin.* 1993;2:215–227.
2. Myers J, Froelicher VF. Exercise testing: procedures and implementation. *Cardiol Clin.* 1993;2:199–214.
3. Pina IL, Balady GJ, Hanson P, et al. Guidelines for clinical exercise testing laboratories: a statement for healthcare professionals from the committee on exercise and cardiac rehabilitation, American Heart Association. *Circulation.* 1995;91:912–921.
4. Strauss HW, Miller DD, Wittry MD, et al. Procedure guidelines for myocardial perfusion imaging. *J Nucl Med.* 1998;39:918–923.
5. Schlant RC, Blomquist CG, Brandenburg RO, et al. Guidelines for exercise testing: a report of the American College of Cardiology/American Heart Association Task Force on Assessment of Cardiovascular Procedures Subcommittee on Exercise Testing. *J Am Coll Cardiol.* 1986;8:725–738.
6. Garcia EV. Imaging guidelines for nuclear cardiology procedures. Part I. *J Nucl Cardiol.* 1996;3:G3–G45.
7. Everaert H, Maes A, Hambey AS, et al. Nuclear cardiology. Part III. Scintigraphic evaluation of cardiac perfusion. *J Nucl Med Technol.* 1998;26:164–169.

Chapter 4

Pharmacologic Stress

Physical exercise is the preferred method for evaluating cardiac performance, but not all patients will be able to perform the amount of exercise needed to achieve the desired heart rate for a valid study. Peripheral vascular disease, neurologic problems, disabling arthritis, previous stroke, orthopedic problems, chronic pulmonary disease, extremity amputations, and poor physical conditions are examples of some of the conditions that would limit physical exercise. In addition, patients on beta blocker medications will not be able to achieve the desired heart rate with exercise stress. Pharmacologic stress is an alternative to physical exercise. Pharmacologic stress can induce the disparity in blood supply between ischemic and nonischemic vessels that would be induced by adequate physical exercise.

Pharmacologic Stress Agents

Several pharmacologic stress agents are used in myocardial perfusion imaging (MPI). The most commonly used pharmacologic agents are dipyridamole, adenosine, and dobutamine. Arbutamine was recently approved for routine use in the United States. These agents fall into two categories: coronary vasodilating agents and cardiac positive inotropic and chronotropic agents (Table 4.1). Vasodilating agents work directly on the coronary vessels to increase blood flow. Inotropic agents work indirectly by increasing myocardial workload, which then leads to an increase in coronary blood flow. Both classes of agents have high accuracy for diagnosing coronary artery disease (CAD) and excellent safety records with acceptably low occurrences of side effects (*1*).

Each agent, however, can cause cardiac and noncardiac side effects. The pharmacologic stress agents and their inhibitors should be administered only by qualified medical personnel (*2*). Cardiac emergency equipment and drugs must be present as they would be for exercise stress testing. An informed consent should be obtained before any pharmacologic stress procedure.

Dipyridamole

Pharmacologic Effects

Dipyridamole produces coronary vasodilation and reactive hyperemia by increasing endogenous plasma adenosine levels. Adenosine is a potent direct coronary artery vasodilator with a very short plasma half-life. Dipyridamole has a plasma half-life of 15–30 minutes. Dipyridamole inhibits the clearance pathway of adenosine across the cell membrane (*3,4*).

Patient Selection

A dipyridamole stress imaging study is indicated for patients who cannot perform exercise stress. Because of the type of side effects that dipyridamole can induce, it is not recommended for patients with acutely unstable angina or recent infarction.

Xanthine-containing medications block the effect of dipyridamole (and adenosine). The need to stop administration of xanthine derivatives before testing is typically not a problem for most outpatients, but the referring physician should be consulted before this type of medication is discontinued (*3*). Bronchodilators are xanthine-containing medications and should be discontinued only under medical supervision. Exacerbation of bronchospasm has been reported in patients with obstructive lung disease upon withdrawal of theophylline-containing bronchodilators (*5*).

Other cardiac medications, such as beta blockers, digitalis, nitrates, or oral dipyridamole, do not interfere with this pharmacologic stress agent.

Patient Preparation and Instructions

The patient should have nothing by mouth (NPO) for 4–6 hours before dipyridamole infusion (similar to preparation for exercise stress testing) to avoid nausea and vomiting. Xanthine-containing medications (such as theophylline) should be stopped 36 hours before the test. Patients should be instructed to have nothing with caffeine for 24 hours before the test, because caffeine blocks the effect of dipyridamole. Common caffeine-containing sustances are listed in Table 4.2. It is important for technologists or other persons taking a history to check for over-the-counter medications that may contain caffeine (headache pills) or xanthines (asthma pills) (*4*).

Dose and Procedure

Before approval for intravenous administration, dipyridamole was administered orally for myocardial perfusion

TABLE 4.1

Comparison of Pharmacologic Stress Agents

	Agents	Patient preparation	Dose	Common side effects (% of patients experiencing)	Contraindications
Coronary vasodilators	Dipyridamole	NPO 4–6 h Xanthine-containing drugs held 36 h Caffeine held 24 h	0.57 mg/kg infused over 4 min	Chest pain (19.7%) Headache (12.2%) Dizziness (11.8%) Ischemic ST segment changes (7.5%) Hypertension (4.6%) Nausea (4.6%) Other—see package insert	Bronchospasm or severe obstructive lung disease Unstable angina Recent myocardial infarction Hypotension (systolic BP < 90) AV block without pacemaker Xanthine-containing drugs Caffeine consumption in last 24 h Allergy to dipyridamole or adenosine
	Adenosine	Same as for dipyridamole Oral dipyridamole held for 24 h	140 µg/kg/min for 6 min	Flushing (44%) Chest discomfort (40%) Dyspnea (28%) ST segment depression (3%) Dizziness (12%) Gastrointestinal discomfort (13%) Hypotension (2%) Headache (18%) Throat, neck, and jaw discomfort (15%) Other—see package insert	
Cardiac positive intropic and chronotropic agents	Dobutamine	NPO 4 h Beta blocker medications held 48 h, after careful assessment of the patient by a physician	5 µg/kg/min for 3 min followed by 10, 20, 30, 40 µg/kg/min each consecutive 3 min as tolerated by patient	Chest pain (30%) Cardiac arrhythmia (19%) Headache (14%) Dyspnea (13%) Flushing (12%) Nausea (8%) Other–see references *1,12,15*	Severe hypertension (BP 90/60) Recent myocardial infarction Unstable angina
	Arbutamine		Individualized to patient's heart rate response	Similar to dobutamine	

TABLE 4.2
Common Caffeine-Containing Substances*

Coffee: brewed and instant
Tea: brewed, instant, and canned
Colas and other soft drinks
All chocolate
Cocoa

* Note: decaffeinated products also contain some caffeine

stress imaging studies. The dose was approximately 375–400 mg. It was assumed that the effect of the drug occurred 45 minutes after ingestion. Because peak effect and results are more variable with oral administration, dipyridamole is no longer used orally for pharmacologic stress with MPI (*6*).

Dipyridamole for intravenous use is diluted with normal saline or 5% dextrose to at least a 1:2 ratio. A volume of 30–50 mL is desired. The dose is 0.57 mg/kg body weight, which is given by intravenous infusion over a 4-minute period of time. The antecubital vein is preferable over the smaller hand veins, which are more sensitive (painful) to the acidic pH of dipyridamole (*3*). The infusion can be performed manually or with a mechanical pump system (although not necessarily).

Before administering the dose, the patient is connected to the electrogardiograph (ECG) and a baseline ECG, blood pressure, and pulse are taken. The ECG should be recorded continuously. Blood pressure and heart rate should be monitored every 1–2 minutes for the first 15 minutes and then, as needed, up to 30 minutes.

The radiopharmaceutical is injected 3–5 minutes after completion of the dipyridamole injection. This allows time to achieve maximum hyperemia of the coronary circulation and the best chances of obtaining an optimal diagnostic study (Fig. 4.1). In the 3–5-minute waiting period after the completion of dipyridamole infusion and radiopharmaceutical administration, the patient can be asked to conduct some low-level exercise, such as handgrip exercise or walking in place. These interventions should increase coronary flow and improve tracer uptake and image quality (*1,7,8*).

Side Effects and Antidote

Severe side effects from dipyridamole are rare. The incidence of life-threatening side effects is similar to that with exercise testing. The most common side effects reported include chest pain, headache, dizziness, and ischemic ST changes. Other less frequently reported side effects have included hypotension, nausea, flushing, dyspnea, and other ECG abnormalities.

The antidote aminophylline should always be kept on hand and ready to administer as needed. Between 10% and 30% of patients treated with dipyridamole will require aminophylline. Aminophylline (theophylline) blocks the effects of adenosine at the receptor site in the vessel and effectively reverses the adverse effects. The dose is generally 50–75 mg, which is given as a bolus. If needed, a second comparable bolus dose or intravenous infusion of 250–500 mg can be given 20 minutes later. On occasion, nitroglycerin also will be needed for angina and/or ST segment

Figure 4.1

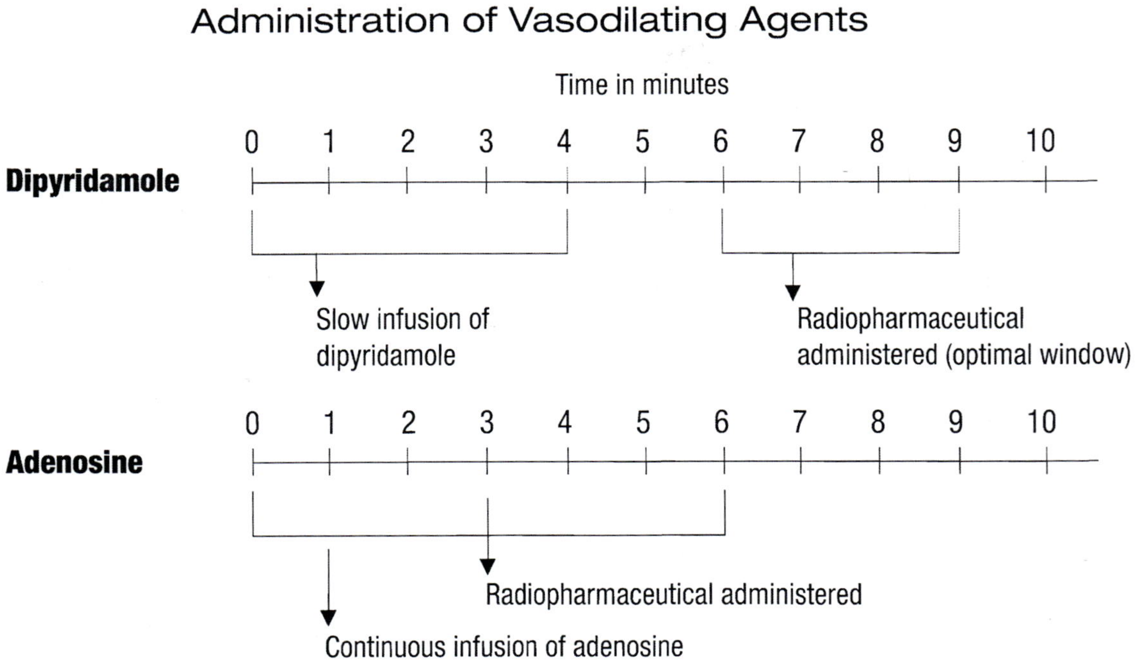

depression not relieved by aminophylline. If any symptoms persist, the patient should be closely monitored and other interventions utilized as required.

The radiopharmaceutical should be administered and allowed at least 1 minute to circulate before administration of aminophylline. The 1-minute delay allows for the maximum uptake of the radiopharmaceutical by the myocardium. If aminophylline is administered before or shortly after (30–60 s) the radiopharmaceutical, it may cause an immediate reversal of the vasodilation, resulting in a nondiagnostic study.

Some physicians administer aminophylline and/or caffeine as a precautionary measure to all patients before releasing them, regardless of symptoms. Some physicians do this selectively.

Contraindications

The contraindications to dipyridamole stress testing include bronchospastic or severe obstructive lung disease (such as with acute asthma or severe chronic obstructive pulmonary disorder), unstable angina, recent myocardial infarction, hypotension (systolic blood pressure <90 mm Hg), greater than first-degree AV block without a pacemaker, patients on xanthine-containing drugs, known allergy to dipyridamole or adenosine, and caffeine consumption in the most recent 24-hour period (*7,9*).

Adenosine

Pharmacologic Effects

Adenosine is a naturally occurring potent vasodilator in most vascular beds. The vasodilatory effect of adenosine is a complex process, the exact mechanism of which is not completely understood (*10*). Adenosine is rapidly cleared from the circulation primarily by erythrocytes and vascular endothelial cells. Its pharmacologic effect is extremely rapid, with a plasma half-life <10 seconds. In the presence of atherosclerotic CAD, the vasodilatory effect of adenosine is blunted. Stenotic coronary vessels unable to dilate will receive a disproportionately decreased blood flow.

Patient Selection

Patient selection for adenosine stress testing is the same as that for the dipyridamole stress test.

Patient Preparation and Instructions

Refer to the patient preparation and instructions for dipyridamole, which are the same for adenosine. Also, oral dipyridamole should be withheld for 24 hours before adenosine pharmacologic stress. Patients should be instructed about the side effects they are likely to experience. They can be told that the adverse effects usually resolve quickly and spontaneously after the infusion is stopped.

Dose and Procedure

Adenosine is administered by intravenous infusion at a rate of 140 µg/kg/min continuously for 6 minutes. The radiopharmaceutical is administered (slowly over 4–5 s) 3 minutes after the start of the adenosine infusion. Adenosine is supplied in a concentration of 3 mg/mL. The total volume needed for a specific patient is calculated as:

$$\text{Total volume in mL} = 0.28 \times \text{total body weight in kg.}$$

The appropriate infusion rate is expressed as:

$$\text{Infusion rate (mL/min)} = \frac{0.140\ \text{(mg/kg/min)} \times \text{total body weight (kg)}}{\text{adenosine concentration (3 mg/mL)}}.$$

Side Effects and Antidote

Side effects with adenosine are more frequent than with dipyridamole, but, because of the short half-life, they can be reversed immediately by terminating the infusion. Common side effects include flushing, chest pain, dyspnea, ECG ST segment depression, dizziness, nausea, and hypotension. Other reported side effects include headache and neck, throat, and jaw discomfort. A high degree of atrioventricular block, although rare and transient, is the most adverse side effect of adenosine infusion (*10*).

Contraindications

The contraindications for adenosine stress testing are the same as those for dipyridamole.

Dobutamine — BEST FOR COPD

Pharmacologic Effects

Dobutamine is a drug that increases heart rate and myocardial contractility. At sufficiently high doses it also increases systolic blood pressure. This will increase the demand for myocardial oxygen consumption. As a result, normal coronary arteries will dilate (*11*).

Patient Selection

Dobutamine pharmacologic stress is indicated for patients who cannot perform treadmill exercise and have contraindications for dipyridamole and adenosine (*12*).

Patient Preparation and Instructions

The patient should be fasting for a minimum of 4 hours. Beta blocker medications should be discontinued as with exercise testing and after careful assessment of the patient.

Dose and Procedure

A baseline ECG, blood pressure, and pulse should be taken before dobutamine administration. Dobutamine is administered intravenously by an infusion pump. The first dose administered is 5 mg/kg/min for 3 minutes followed by 10, 20, 30, and 40 mg/kg/min for each consecutive 3-minute period as tolerated by the patient. The patient's ECG should be monitored throughout the infusion period. Blood pressure and pulse should be recorded every minute throughout the infusion. The patient should be monitored after the infusion is terminated until symptoms subside and the baseline values of all parameters are achieved.

The radiopharmaceutical should be administered 1 minute after initiation of the maximum tolerated dose of dobutamine. The infusion at that rate should continue for another 2 minutes. The maximum tolerated dose is that which precipitates ischemic symptoms.

Dobutamine infusion may be supplemented if needed with a dose of 0.5–1 mg atropine administered intravenously to achieve the desired heart rate (85% maximum predicted heart rate) (*1,13,14*).

Side Effects and Antidote

Seventy-five percent or more of patients experience side effects during dobutamine infusion. Reported side effects include chest pain, cardiac arrhythmia, headache, dyspnea, flushing, and nausea. Because the tissue half-life of dobutamine is short lived, drug-related side effects usually resolve within 5–10 minutes after discontinuation of the infusion (*1,12,15*).

Contraindications

Dobutamine injection is contraindicated in patients with idiopathic hypertrophic subaortic stenosis and in patients who have shown previous manifestations of hypersensivity to dobutamine.

Relative contraindications to dobutamine stress testing include recent myocardial infarction or unstable angina, severe hypertension (blood pressure 180/110), or hypotension (blood pressure 90/60) (*12,13*).

Arbutamine

Arbutamine is a synthetic catecholamine and an inotropic as well as chronotropic agent. Arbutamine simulates the cardiac response of exercise by increasing heart rate and myocardial contractility. It is administered by a computer-controlled heart rate feedback (closed loop) delivery system that monitors the response to the arbutamine infusion. Patient selection for arbutamine stress is similar to that for dobutamine. Side effects with arbutamine administration are also similar to those with dobutamine (*16,17*).

Documentation

It is useful to document the type of pharmacologic stress used so that this information is readily available to the physician who will read and interpret the study. Sample consent forms and a sample form for documentation are included in Figures 4.2–4.5.

Alternatives to Treadmill Exercise and Pharmacologic Stress

A small percentage of patients will not be able to perform treadmill exercise or undergo a pharmacologic stress test. Other methods to evaluate these patients for CAD have been investigated. They include the cold pressor test, in which the hands are immersed in ice water, isometric handgrip exercise, and mental stress. Each of these methods can cause some increase in heart rate and blood pressure (*18,19*). They are inadequate at inducing the ischemic conditions, chest pain, and ST segment changes needed to evaluate CAD with MPI and are not recommended alternatives to pharmacologic or physical stress.

Atrial Pacing

Transesophageal atrial pacing has been used on occasion to increase heart rate to 85% of the maximum predicted value and induce symptoms of angina (*20,23*). A myocardial perfusion agent can be injected at the appropriate time during atrial pacing. The electrode is introduced through the nose or is swallowed as a pill containing the electrode. The procedure is somewhat uncomfortable for patients and, because pharmacologic stress agents are widely available, is rarely indicated.

Figure 4.2 Sample consent for intravenous dipyridamole nuclear stress test.

Consent for Intravenous Dipyridamole Nuclear Stress Test

I,______________________________, authorize the physicians, cardiac clinical specialist, and technologists of the______________________________ to administer and conduct the intravenous (IV) dipyridamole test, which was requested by my physician.

This test is designated to detect and/or evaluate the presence or absence of significant heart disease. I understand that this test uses a coronary vessel dilator (dipyridamole) and the results may be of value to my physicians in deciding the best course of treatment for me.

During the performance of this test, my electrocardiogram will be monitored continuously and my blood pressure and pulse rate will be measured and recorded at periodic intervals.

It is fairly common that mild symptoms of chest pain, headache, nausea, or flushed sensations may occur. I also understand that on extremely rare occasions a heart attack (myocardial infarction), fainting, or abnormal heart rhythms may occur. I further understand that the laboratory is properly equipped for such situations and that its professional personnel are trained to administer any emergency care necessary.

I also understand that I may request that the test be discontinued at any time.

I understand this test will be conducted by a physician and/or cardiac clinical specialist nurse and expert technologist who will conduct this test in a careful manner. Every effort will be made to minimize discomfort and risk.

The test will be discontinued or reversed by the administration of IV aminophylline when indicated.

In addition, thallium-201 or technetium-99m (radioactive elements in solution for the purpose of heart scanning) will be administered into a vein as part of this test. There are no known side effects from this administration.

I understand the above information and give my full and informed consent to proceed with the IV dipyridamole test.

______________________________ ______________________________

Signature of patient Date Signature of witness Date

Figure 4.3 Sample consent for intravenous adenosine nuclear stress test.

Consent for Intravenous Adenosine Nuclear Stress Test

I, ______________________________, authorize the physicians, cardiac clinical specialist, and technologists of ______________________________ to administer and conduct the intravenous (IV) adenosine test, which was requested by my physician.

This test is designated to detect and/or evaluate the presence or absence of significant heart disease. I understand that this test uses a coronary vessel dilator (adenosine) and the results may be of value to my physicians in deciding the best course of treatment for me.

During the performance of this test, my electrocardiogram will be monitored continuously and my blood pressure and pulse rate will be measured and recorded at periodic intervals.

It is fairly common that mild symptoms of chest pain, headache, nausea, or flushed sensations may occur. I also understand that on extremely rare occasions a heart attack (myocardial infarction), fainting, or abnormal heart rhythms may occur. I further understand that the laboratory is properly equipped for such situations and that its professional personnel are trained to administer any emergency care necessary.

I also understand that I may request that the test be discontinued at any time.

I understand that a physician and/or cardiac clinical specialist nurse and expert technologist will conduct this test in a careful manner. Every effort will be made to minimize discomfort and risk.

In addition, technetium-99m (a radioactive element in solution for the purpose of heart scanning) will be administered into a vein as part of this test. There are no known side effects from its administration.

I understand the above information and give my full and informed consent to proceed with the IV adenosine test.

I attest that if I am a woman of childbearing age, I am not pregnant or breastfeeding at this present time.

Signature of patient	Date	Signature of witness	Date

Figure 4.4 Sample consent for dobutamine stress test.

Consent for Dobutamine Stress Test

To facilitate the diagnosis and appropriate management of my condition, I hereby voluntarily consent to engage in a dobutamine stress test to help evaluate the functional state of my heart and circulation. Before undergoing the test, I shall be interviewed by a physician in order to determine if I have any condition that would indicate that I should not engage in this test.

I understand that I shall receive progressively higher doses of a medicine, dobutamine, which somewhat simulates exercise. During the test, my electrocardiogram will be continuously monitored and my blood pressure periodically recorded. The medication will be discontinued when I attain a predetermined end point, develop any abnormal change that the physician feels is significant, or have distressing symptoms, such as chest discomfort, shortness of breath, or fatigue.

I understand that every effort will be made to conduct the dobutamine stress test in such a way as to minimize discomfort or risk, although such tests are occasionally associated with certain changes during or just after the procedure. These include episodes of flushing, transient lightheadedness, fainting, chest discomfort, changes in the heart rhythm, and, in exceptionally rare instances, heart attacks. I understand that the laboratory is properly equipped with emergency equipment and that trained personnel are available to deal with any unusual situations that may arise.

I have read the foregoing and understand it. Any questions that may have occurred to me have been answered to my satisfaction.

______________________________ ______________________________

Signature of patient Date Signature of witness Date

Figure 4.5 Sample pharmacologic stress documentation form.

Pharmacologic Stress

Persantine **Dobutamine** **Adenosine**

Patient ____________________

Date of test ____________________ Time ____________________

Physician ____________________ Nurse/tech ____________________

Referring physician ____________________ Protocol ____________________

Medications ____________________

____

Persantine Results

_______ mg of dipyridamole were given intravenously over 4 min.

(SestaMIBI) (Thallium) was given at _______ minutes by the nuclear medicine tech.

_______ mg of aminophylline were given intravenously at _______ minutes to reverse the effects of the dipyridamole.

Dobutamine Results

Dobutamine was infused at increasing doses to a maximum of _______ mg/kg/min.

_______ mg of atropine were given to reverse the effects of the dobutamine.

Adenosine Results

_______ mg of adenosine were given over 6 minutes.

(SestaMIBI) (Thallium) was given at 3 minutes from start of adenosine infusion.

Interpretation:

Normal Abnormal Borderline Inconclusive

References

1. Travin MI, Wexler JP. Pharmacological stress testing. *Semin Nucl Med.* 1999;29:298–318.
2. Everaert H, Maes A, Hambye AS, Mesotten L, Mortelmans L, Franken PR. Nuclear cardiology. Part III. Scintigraphic evaluation of cardiac perfusion. *J Nucl Med Technol.* 1998;26:164–169.
3. Leppo JA. Dipyridamole–thallium imaging: the lazy man's stress test. *J Nucl Med.* 1989;30:281–287.
4. Blust JS, Boyce TM, Moore WH. Pharmacological cardiac intervention: comparison of adenosine, dipyridamole, and dobutamine. *J Nucl Med Technol.* 1992;20:53–59.
5. Homma S, Gilliland Y, Guiney TE, Strauss HW, Boucher CA. Safety of intravenous dipyridamole for stress testing with thallium imaging. *Am J Cardiol.* 1987;59:152–154.
6. Iskandrian AS. Single-photon emission computed tomographic thallium imaging with adenosine, dipyridamole and exercise. *Am Heart J.* 1991;122:279–284.
7. Leppo JA. Dipyridamole myocardial perfusion imaging. *J Nucl Med.* 1994;35:730–733.
8. Strauss HW, Miller DD, Wittry MD, et al. Procedure guideline for myocardial perfusion imaging. *J Nucl Med.* 1998;39:918–923.
9. Garcia EV. Imaging guidelines for nuclear cardiology procedures. Part I. *J Nucl Cardiol.* 1996;3:G3–G45.
10. Iskandrian AS. Adenosine myocardial perfusion imaging. *J Nucl Med.* 1994;35:734–736.
11. Verani MS. Dobutamine myocardial perfusion imaging. *J Nucl Med.* 1994;35:737–739.
12. Hays JT, Mahmarian JJ, Cochran AJ, Verani MS. Dobutamine thallium-201 tomography for evaluating patients with suspected coronary artery disease unable to undergo exercise or vasodilator pharmacological stress testing. *J Am Coll Cardiol.* 1993;21:1583–1590.
13. Elhendy A, van Domburg RT, Bax JJ, et al. Dobutamine-atropine stress myocardial perfusion SPECT imaging in the diagnosis of graft stenosis after coronary artery bypass grafting. *J Nucl Cardiol.* 1998;5:491–497.
14. Huang PJ, Yen RF, Chieng PU, Chen ML, Su CT. Do beta-blockers affect the diagnostic sensitivity of dobutamine stress thallium-201 single photon emission computed tomographic imaging? *J Nucl Cardiol.* 1998;5:34–39.
15. Dakik HA, Vempathy H, Verani MS. Tolerance, hemodynamic changes and safety of dobutamine stress perfusion imaging. *J Nucl Cardiol.* 1996;3:410–414.
16. Anagnostopoulos C, Pennell D, Francis J, Serup-Hansen K, Davies G, Underwood R. A comparison of adenosine and arbutamine for myocardial perfusion imaging. *Eur J Nucl Med.* 1998;25:394–400.
17. Dennis CA, Pool PE, Perrins EJ, et al. Stress testing with closed-loop arbutamine as an alternative to exercise. *J Am Coll Cardiol.* 1995;26:1151–1158.
18. Northcote RJ, Cooke MB. How useful are the cold pressor test and sustained isometric handgrip exercise with radionuclide ventriculography in the evaluation of patients with coronary artery disease? *Br Heart J.* 1987;57:319–328.
19. Schoder H, Silverman DH, Campisi R, et al. Effect of mental stress on myocardial blood flow and vasomotion in patients with coronary artery disease. *J Nucl Med.* 2000;41:11–16.
20. Le Feuvre C, Vacheron A, Metzger JP, et al. Prognostic value of thallium-201 myocardial scintigraphy after atrial transoesophageal pacing in patients with suspected coronary artery disease. *Eur Heart J.* 1993;14:1195–1199.
21. Le Feuvre C, Vacheron A, Metzger JP, et al. Comparison of thallium myocardial scintigraphy after exercise and transoesophageal atrial pacing in the diagnosis of coronary artery disease. *Eur Heart J.* 1992;13:794–800.
22. Cuocolo A, Santomauro M, Pace L, et al. Comparison between exercise and transeosophageal atrial pacing in patients with coronary artery disease: technetium 99m methoxy isobutyl isonitrile simultaneous evaluation of ventricular function and myocardial perfusion. *Eur J Nucl Med.* 1992;19:119–124.
23. Stratmann HG, Mark AL, Walker KE, Williams GA. Prognostic value of atrial pacing and thallium-201 scintigraphy in patients with stable chest pain. *Am J Cardiol.* 1989;64:985–990.

Chapter 5

Thallium Imaging Procedure

Planar and/or SPECT imaging can be acquired with thallium or with any ^{99m}Tc-labeled radiopharmaceutical. SPECT imaging should be routinely performed because it significantly improves sensitivity of the test and is better for assessing the extent of disease (*1,2*). SPECT images can be gated to evaluate wall motion and left ventricular ejection fraction. Planar imaging is particularly useful for evaluating the heart-to-lung ratio of thallium and breast or diaphragmatic attenuation. When planar imaging is performed without SPECT, a minimum of three views is required. Those views are the anterior, left anterior oblique (LAO) angled for best septal visualization, and the left lateral or 70° LAO. Patient positioning and data acquisition parameters for planar and SPECT image acquisition and gated imaging are described in greater detail in chapter 9.

Sequence of Imaging Studies

Stress Study

When thallium alone is used for myocardial perfusion imaging (MPI), the stress study is always performed first. The stress can be physical exercise or pharmacologic. Imaging should begin as soon as possible (preferably within 5–7 minutes) after the radiopharmaceutical is injected. The stress lab should be in close proximity to the camera room. Because the patient is quickly transferred from the stress lab to the imaging area, his or her condition should be closely monitored to ensure comfort and safety. The stress images should be acquired as quickly as possible to minimize the effects of thallium redistribution.

Dose

For the stress study, 2–4 mCi thallium are administered. The maximum dose of thallium for MPI is 4 mCi. When a second dose of thallium at rest is likely to be given, 2.5–3 mCi are administered for the stress study and 1.0–1.5 mCi are used for the second injection at rest.

Three- to Four-Hour Delayed (Redistribution) Study

A second set of images, the delayed rest and redistribution images, can be acquired 3–4 hours after the injection. The delayed images will be compared with the stress images. The acquisition protocol and patient positioning for the two studies should be exactly the same.

During the waiting period between studies, the thallium may redistribute into areas of the myocardium that were hypoperfused at stress. Eating between the stress and delayed thallium studies may have a negative effect on the myocardial redistribution of thallium and therefore should be avoided (*3*). During the waiting period between the stress and delayed study, the patient should drink only clear liquids, such as coffee and juice. Physical activity also should be minimized. The patient should sit and rest. Activity, such as walking, shopping, and climbing stairs should be discouraged.

Scan Pattern Interpretation

If the uptake of radiopharmaceutical is uniform throughout the myocardium on both studies and there are no regions of hypoperfusion, then the study is essentially negative for coronary artery disease. If a region of the myocardium is hypoperfused on the stress study and shows significant improvement on the delayed study, a region of transient ischemia has been identified. A region of hypoperfusion on the stress study that shows little or no improvement on the 4-hour delayed study is referred to as a fixed perfusion deficit.

A fixed perfusion deficit requires further investigation. This region may be chronic ischemia, hibernating myocardium, or scarred tissue from a previous myocardial infarction (MI). Regions of hypoperfusion on a stress study that are nonreversible on the 3–4-hour delayed study will sometimes show improvement by 18–24 hours after the thallium injection, which is usually the case in hibernating but viable myocardium. The rate at which thallium equilibrates varies, depending on several factors, such as plasma concentration of thallium and regional blood flow. As discussed in chapter 2, 4-hour redistribution imaging after a single dose of thallium given at stress can be grossly inaccurate in distinguishing a region of ischemic but viable myocardium from nonviable myocardium (*4*).

Twenty-Four-Hour Delayed (Redistribution) Scan

When a stress and 3–4-hour delayed imaging study performed with a single dose of thallium given at stress reveals a fixed perfusion defect, the patient could be asked to return the following day for another set of images. These images would be compared with the previous images to see if the region of hypoperfusion reverses (improves) or not.

TABLE 5.1

Thallium Imaging Procedure Options

Procedure 1

1. Administer 4 mCi ^{201}Tl at peak stress.
2. Acquire a stress scan immediately after the ^{201}Tl injection (beginning within 5–7 minutes).
3. Acquire a 3–4-hour delayed (redistribution) scan.
4. When a fixed defect is identified, acquire a 24-hour delayed (redistribution) scan.

Procedure 2

1. Administer 2.5–3.0 mCi ^{201}Tl at peak stress.
2. Acquire a stress scan immediately after the ^{201}Tl injection (beginning within 5–7 minutes).
3. Acquire a 3–4-hour delayed (redistribution) scan.
4. When a fixed defect is identified, administer 1.0–1.5 mCi ^{201}Tl the same day.
5. Perform a rest scan 10 minutes later.

Procedure 3

1. Administer 4 mCi ^{201}Tl at peak stress.
2. Acquire a stress scan immediately after the ^{201}Tl injection (beginning within 5–7 minutes).
3. Acquire a 3–4-hour delayed (redistribution) scan.
4. When a fixed defect is identified, administer 3–4 mCi ^{201}Tl on a subsequent day.
5. Perform a rest scan 10 minutes later.

Procedure 4

1. Administer 2.5–3.0 mCi ^{201}Tl at peak stress.
2. Acquire a stress scan immediately after the ^{201}Tl injection (beginning within 5–7 minutes).
3. Immediately after completion of the stress scan, administer 1.0–1.5 mCi ^{201}Tl.
4. Acquire a rest scan 3–4 hours later.

A significant number of fixed defects on a 3–4-hour delayed study will demonstrate late reversibility (*5*). Myocardial segments that demonstrate late reversibility are ischemic but viable, similar to ischemic regions identified on a 3–4-hour delayed study (*6*). Revascularization procedures will improve the status of these ischemic areas. Problems associated with the performance of a 24-hour delayed study include the scheduling of camera time, patient compliance with returning the next day, and a lower count density of thallium as a result of physical decay.

Second Thallium Injection Technique

Another approach to resolving the viability question of the fixed defects observed on the 3–4-hour thallium delayed image study is to administer a second dose of thallium at rest and then reimage. The second dose of thallium could be given immediately after the 3–4-hour delayed scan. This is a selective approach based on the scan pattern of the first two imaging studies. The images can be acquired the same day, 10 minutes after the second injection (*7,8,9*).

The same-day reinjection imaging technique is a practical and convenient alternative to 24-hour imaging. A reinjection method is also considered to be more accurate at determining viability than the single-injection procedure. With the single-injection procedure, the stress images represent initial blood flow and the delayed images show reperfusion—two different phenomena. With the reinjection technique, there will be two sets of images reflecting blood flow (*7*).

When a second injection is given on the same day as the initial injection for the stress study, the 4-mCi dose must be split as previously described. If a second injection is given on a subsequent day, up to 4 mCi can be given for both injections (*9,10*).

A second injection of thallium also could be administered immediately after the completion of the stress imaging study. A resting scan could then be acquired 3–4 hours later. This same-day procedure would require the least amounts of camera time and patient cooperation. The patient should avoid eating a meal between stress and 3–4-hour delayed redistribution and equilibrium images.

Table 5.1 lists four possible thallium imaging procedures. The purpose, advantages, and disadvantages of the different procedures have been described. The first three procedures require two to three sets of images. The fourth procedure requires only two sets of images. Table 5.2 describes the clinical interpretation of various scan patterns.

Limitations of Thallium in the Detection of Coronary Artery Disease

The major limitations of thallium as an ideal agent for MPI are physical characteristics and radiation dosimetry. The low energy of the principle photons (68–80 keV) from thallium results in significant soft tissue attenuation and scatter. Thallium is not an optimal agent to use for obese patients. Care should be taken to avoid false-positive interpretation as a result of diaphragmatic attenuation (more commonly seen in men) and breast attenuation (more commonly seen in females).

A false-negative study can result from triple-vessel coronary artery disease. Triple-vessel disease may cause uniform decreased uptake in all segments of the myocardium. Increased lung uptake of thallium and/or left ventricular cavity dilatation immediately after stress are associated with extensive disease. Both of these can be detected visually by

TABLE 5.2
Clinical Significance of Various Scan Patterns

Stress study	3–4-hour delayed study	24-hour delayed and/or rest study	Clinical interpretation
Normal	Normal	–	Coronary artery disease unlikely
Abnormal	Normal	–	Stress-induced ischemia
Abnormal	Abnormal	Normal	Chronic ischemia
Abnormal	Abnormal	Abnormal	Fixed defect, nonviable, or scar from previous myocardial infarction

reviewing an anterior planar view taken immediately after exercise. A delayed planar anterior image should be acquired just before the delayed SPECT study for comparison with the stress planar image.

Heart–Lung Ratio of Thallium

Quantitation of heart-to-lung activity minimizes subjectivity in the evaluation of left ventricular dysfunction. Quantitation of the lung-to-heart activity can be made by comparing the counts in a region of interest (ROI) in the lungs with an ROI in the myocardium. Both ROIs should be the same size (same number of pixels). This is generally done on poststress anterior planar images. Normally, the lung-to-heart ratio is between 0.3 and 0.5.

Increased pulmonary uptake immediately after stress is related to elevated left ventricular pressure. Conditions that can cause elevated left ventricular pressure include severe CAD, hypertension, and valvular disease.

References

1. Beller G. Myocardial perfusion imaging with thallium-201. *J Nucl Med.* 1994;35:674–680.
2. Kiat H, Berman D, Maddahi J. Comparison of planar and tomographic exercise thallium-201 imaging methods for the evaluation of coronary artery disease. *J Am Coll Cardiol.* 1989;13:613–616.
3. Angello D, Wilson R, Palac R. Effects of eating on thallium-201 myocardial redistribution after myocardial ischemia. *Am J Cardiol.* 1987;60:528–533.
4. Kiat H, Berman D, Maddahi J, et al. Late reversibility of tomographic myocardial thallium-201 defects: an accurate marker of myocardial viability. *J Am Coll Cardiol.* 1988;12:1456–1463.
5. Yang L, Berman D, Kiat H, et al. The frequency of late reversibility in SPECT thallium-201 stress redistribution studies. *J Am Coll Cardiol.* 1990;15:334–340.
6. Botvinick E. Late reversibility: a viability issue. *J Am Coll Cardiol.* 1990;15:341–344.
7. Pieri P, McKusick K, Fischman A, Alpert N, Bougas C, Strauss HW. Characterization of the reinjection thallium image. *J Nucl Med.* 1993;34:879–884.
8. Ohtani H, Tamaki N, Yonekura Y, et al. Value of thallium-201 reinjection after delayed SPECT imaging for predicting reversible ischemia after coronary artery bypass grafting. *Am J Cardiol.* 1990;66:394–399.
9. Inglese E, Brambilla M, Dondi M, et al. Assessment of myocardial viability after thallium-201 reinjection or rest redistribution imaging: a multi-center study. *J Nucl Med.* 1995;36:555–563.
10. Strauss HW, Miller DD, Wittry MD, et al. Procedure guideline for myocardial perfusion imaging. *J Nucl Med.* 1998;39:918–922.

Chapter 6

Sestamibi and Tetrofosmin Imaging Procedures

^{99m}Tc-sestamibi and ^{99m}Tc-tetrofosmin do not redistribute as does thallium and, for the most part, remain fixed in the myocardium. Two separate injections are required, one for the rest study and one for the stress study. Either the rest or stress study can be performed first. Both stress and rest studies can be completed in 1 day or could be performed on separate days.

Technetium-99m–Sestamibi and Technetium-99m–Tetrofosmin

Dose

The recommended dose for sestamibi ranges from 10–30 mCi and for tetrofosmin from 5–24 mCi. These doses vary, depending on whether a one- or two-day protocol is used. When a one-day protocol is used, a smaller dose is administered for the first imaging study and a higher dose is administered for the second study. If the rest and stress studies are performed on 2 separate days, a larger dose usually is administered for each study (Table 6.1).

Effect of Extra Cardiac Activity

A significant amount of each agent accumulates in the liver and gallbladder and is then excreted via the bowel. This results in subdiaphragmatic attenuation artifacts, particularly in the inferior wall. The imaging time after injection is selected to minimize interfering liver, gallbladder, and bowel activity. A glass of milk and/or other fatty meal can be used to clear excessive gallbladder activity. The use of milk or fatty meal after injection will result in greater large bowel activity earlier on. Timing of the imaging studies after injection and patient instructions about eating during the waiting period before imaging must take into account liver and bowel activity.

Sestamibi

When sestamibi is injected after exercise, the heart-to-liver ratio is initially 1.0 and improves over time. When sestamibi is injected at rest, the liver concentration is initially almost twice that of the heart. As time progresses, liver activity decreases and gallbladder and bowel activity increase (*1*). If there is interfering gallbladder and bowel activity, the patient can be instructed to take a glass of milk, fatty meal, or a glass of water before imaging. Imaging can begin within 60 minutes after injection. It could also begin as soon as 15–30 minutes after injection to avoid any significant accumulation of subdiaphragmatic activity.

Tetrofosmin

The liver clearance of tetrofosmin is more rapid than that with sestamibi (*2*). As with sestamibi, imaging with tetrofosmin can begin 15–60 minutes after injection.

One-Day Protocol

Several factors should be considered when deciding which study will be performed first (rest or stress) in a same-day protocol. If the stress study is performed first and is normal, the rest study may not be required. Cardiologists may prefer conducting stress tests in the morning. However, the use of a lower dose for the stress study may underestimate defect reversibility (*3,4*). The stress study is the most important in terms of disease detection. When a stress study is the second to be performed, the image statistics are optimized.

Regardless of which study is performed first, a longer imaging time per stop for SPECT acquisition is recommended for the first study. A 3–4-hour delay between studies is also recommended to minimize the influence of counts from the first injection on the second acquisition. The total time for the stress/rest study will take approximately 4–5 hours. Although this is a lengthy procedure for a patient, it eliminates the need to return on a second day and results can be obtained sooner. Figure 6.1 is a diagram of two possible same-day protocols.

TABLE 6.1

Doses of ^{99m}Tc-Labeled Agents for Myocardial Perfusion Scintigraphy

Protocol	Sestamibi	Tetrofosmin
One-day protocol		
First dose	8–10 mCi	5–8 mCi
Second dose	22–30 mCi	15–24 mCi
Two-day protocol		
First dose	15–30 mCi	15–24 mCi
Second dose	15–30 mCi	15–24 mCi

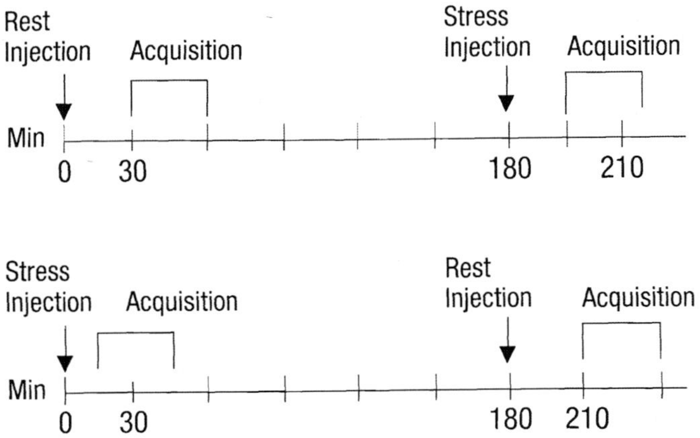

Figure 6.1 Same-day protocols (sestamibi or tetrofosmin).

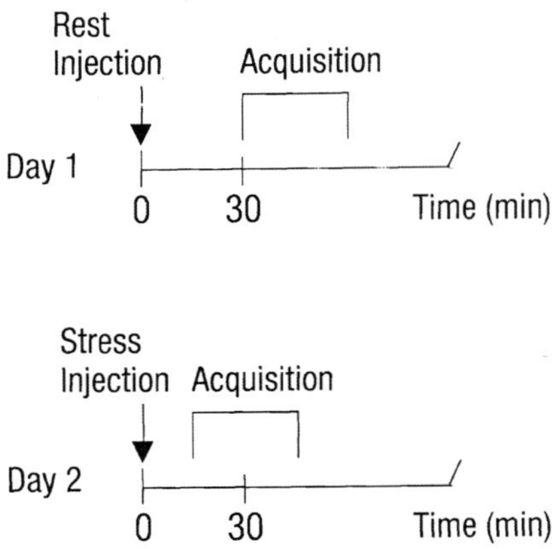

Figure 6.2 Two-day protocol (sestamibi or tetrofosmin). Either stress or rest study can be performed first.

Two-Day Protocol

A large dose is given for both studies when performed on separate days. The advantages of performing the studies on 2 separate days are near complete decay of the radiotracer between studies and optimal statistics from both acquisitions. A two-day study may not be practical for all patients and all settings, particularly if the patient must travel a long distance. For patients who can return on a second day, this protocol allows the maximum flexibility in scheduling for both the patient and clinical site. Figure 6.2 is a diagram of a two-day protocol.

Patient Instructions for Imaging Procedure

With all procedures, patients should be given instructions in writing when the appointment is scheduled. These instructions should include the time and place to report, how long the test will take, dietary and medication instructions, and any other information that will be pertinent to the patient. Figures 6.3 and 6.4 (on the following pages) are sample patient instruction sheets for two-day technetium agent protocols. Each clinic must develop one- and two-day protocol patient instruction sheets that reflect the stress methods and radiopharmaceuticals used.

References

1. Leppo JA, DePuey EG, Johnson LL. A review of cardiac imaging with sestamibi and teboroxime. *J Nucl Med.* 1991;32:2012–2022.
2. Munch G, Neverve J, Matsunari I, Schroter G, Schwaiger M. Myocardial technetium-99m-tetrofosmin and technetium-99m-sestamibi kinetics in normal subjects and patients with coronary artery disease. *J Nucl Med.* 1997;38:428–432.
3. Berman DS, Kiat HS, Van Train KF, Germano G, Maddahi J, Friedman JD. Myocardial perfusion imaging with technetium-99m-sestamibi: comparative analysis of available imaging protocols. *J Nucl Med.* 1994;35:681–688.
4. Jain D. Technetium-99m-labeled myocardial perfusion imaging agents. *Semin Nucl Med.* 1999;29:221–236.

Figure 6.3 Two-day protocol. Patient instructions day 1.

Myocardial Perfusion Stress Imaging Study

Clinic Name
Address

Patient name ______________________________

Stress schedule appointment time ______________________

This is a two-day test. Day 1 involves exercise testing and cardiac imaging. Please be prepared to be here for 90–120 minutes. You will need to return the next business day for resting images of the heart. These images will take approximately 60–90 minutes to complete. It is imperative that these delay images be completed for an accurate exam. Your time to return for resting images will be scheduled at the time of the stress appointment.

Patient Preparation

- Light meal (toast, cereal, juice) prior to 4 hours before your appointment. (Minimum of 3-hour fast before each test, except for patients with diabetes.)
- No caffeine or decaffeinated products for 24 hours before stress imaging.
- Wear loose-fitting clothing and sneakers or comfortable walking shoes.
- Bring a list of your medications, with times and doses indicated.

*** Take all medication unless specified by your doctor.

Please try to arrive about 15 minutes before your scheduled time. **At least 24–48 hours notice must be given if you are unable to keep your appointment.** You may be charged for the price of the radiopharmaceutical if appropriate notice of cancellation is not given.

If you will be undergoing a stress test accompanied with persantine, refrain from taking any xanthine (asthma) medications for 36–48 hours before the test (Aerolate, Lufyllin, Marax tablets, Respbid tablets, Slo-bid Gyrocaps, Theo-extended release tablets, Uni-Dur, Uniphyl).

Please consult your physician before discontinuing any medications.

Special note to women: As with all nuclear medicine exams, if you are pregnant or a nursing mother, this exam is not appropriate at this time unless highly indicated.

If you have any additional questions, please call before your appointment.

Figure 6.4 Two-day protocol. Patient instructions day 2.

Myocardial Perfusion Stress Imaging Study

Clinic Name
Address

Patient name ______________________________

Date and time scheduled for resting scan ______________________________

Now that you have completed the hard part—treadmill or pharmacologic stress—relax. It's all downhill from here!

When you come in, you will receive another injection of sestamibi (the radioactive tracer for the second part of the test). This material is very short-lived, and almost all of it is gone by the second day. Therefore, we have to give you a second dose. This needs to circulate for 30–60 minutes. During the wait, you are encouraged to drink some cold water.

To prepare for this second part, please fast for at least 2 hours before your appointment. Diabetics can eat. (If you are diabetic, please bring some milk with you.) Refrain from drinking coffee for another day. Take all medications as directed. Plan to be here for about 60–90 minutes. Please rest before you come in, because this is a resting scan. Normal daily activities are fine, but try to avoid strenuous activities such as jogging or aerobics.

The results will be ready approximately 1–2 days after you complete both parts of the test procedure. A copy of the results of your test will be sent to your primary doctor as well as your cardiologist.

If you have any additional questions, please feel free to ask.

Chapter 7

Dual-Isotope Technetium-99m and Thallium-201 Imaging

A two-day protocol with a technetium agent will provide an optimal study in most circumstances. ^{99m}Tc has a higher photon energy for imaging than does ^{201}Tl. The larger doses yield better image statistics. The majority of the activity from the first study is decayed within 24 hours. The parameters from the two acquisitions from a two-day technetium protocol are most similar and comparable. Thallium, however, has an ideal property for the assessment of tissue viability. A dual-isotope study utilizes the best properties of each agent and is useful for evaluating myocardial ischemia and viability. Thallium is administered at rest, and a technetium agent is administered for the stress procedure. Images for each study can be acquired separately or simultaneously. It is technically more difficult to perform a simultaneous acquisition of both isotopes because of the contribution of ^{99m}Tc in the ^{201}Tl window.

Separate Acquisition Protocol

The rest study is usually performed first. Three to four millicuries of thallium are administered at rest. The acquisition can begin as soon as 15 minutes after injection. After the completion of the rest study acquisition, a stress study is performed using a technetium-labeled myocardial perfusion imaging agent. Imaging can begin 15 minutes after injection. The images should be reviewed before the patient is released. When comparison of the two studies shows a fixed defect, the patient can be asked to return the next morning for a delayed thallium study. To avoid the possibility of needing three acquisitions over 2 days to complete the study, the thallium injection could be given the night before. The thallium rest–redistribution study can be acquired 12 hours or more later (the next morning), followed by a technetium stress study (*1*). Figure 7.1 represents the dual-isotope imaging protocols.

Simultaneous Acquisition

Significant advantages could accompany the simultaneous acquisition of a thallium rest study and a technetium agent stress study if a reliable method to correct for the contribution to ^{99m}Tc in the ^{201}Tl window could be employed. The total stress–rest procedure could be completed in <1 hour. Camera time would be optimized. The patient's position would be identical for the two studies. Artifacts could be minimized. The reconstructed slices of the two acquisitions would be more closely aligned. However, until a reliable method to correct for the contribution of ^{99m}Tc in the ^{201}Tl window is developed, simultaneous acquisition is not recommended for general use (*2,3*).

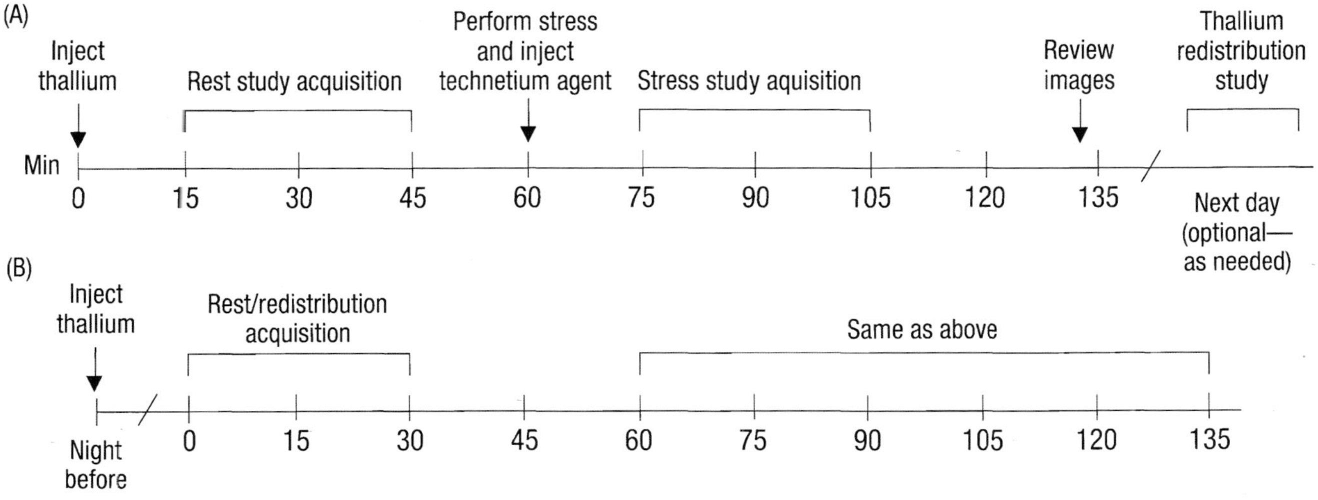

Figure 7.1 Dual-isotope imaging protocols.

References

1. Berman DS, Kiat HS, Van Train KF, et al. Myocardial perfusion imaging with technetium-99m-sestamibi: comparative analysis of available imaging protocols. *J Nucl Med.* 1994;35:681–688.
2. Kiat H, Germano G, Friedman J. Comparative feasibility of separate or simultaneous rest thallium-201/stress technitium-99m-sestamibi dual-isotope myocardial perfusion spect. *J Nucl Med.* 1994;35:542–548.
3. Lowe VJ, Greer KL, Hanson MW, Jaszczak RJ, Coleman RE. Cardiac phantom evaluation of simultaneously acquired dual-isotope rest thallium-201/stress technetium-99m SPECT images. *J Nucl Med.* 1993;34:1998–2006.

Chapter 8

Data Acquisition: Planar and SPECT

Before acquiring patient studies, quality control (QC) studies must be performed on the imaging equipment. QC tests should be performed on a routine basis to detect changes from the baseline condition in an imaging system's performance. Details of data acquisition in planar, SPECT, and gated SPECT imaging are discussed in this chapter.

Camera Quality Control

The baseline condition or parameters of the equipment are established in a process called acceptance testing at the time the equipment is first installed in a facility. Acceptance testing is the first step in the establishment of a QC program. The purpose of acceptance testing is to verify that the equipment performs as specified by the manufacturer according to National Electrical Manufacturers Association (NEMA) guidelines. NEMA has established methods by which performance specifications are defined and measured. The performance measurements for scintillation cameras defined by NEMA are listed in Table 8.1. Acceptance tests are detailed and extensive. They require special phantoms and software not used for routine QC. Other parameters should also be evaluated when a new camera system is installed. Some examples include the alignment of the camera heads with the axis of rotation, alignment of the imaging table parallel to and centered on the axis of rotation, and orientation of reconstructed data. All these tests and measurements should be performed by someone with knowledge of and experience with acceptance testing. For an unbiased evaluation, it is preferable that this person not be a representative of the camera manufacturer (*1,2*).

Once acceptance testing has been completed, a routine camera QC program is established to monitor those performance parameters that can change over time. The importance of a routine QC program cannot be overemphasized. This program should include tests to evaluate the appropriateness of the energy window for the photopeak(s) of the desired radionuclide and the uniformity, resolution, and linearity of the system. These tests are prerequisites to all planar and SPECT imaging studies. The photopeak setting should be checked each day and each time a different radionuclide is to be used. A low count (3–5 million) flood image should be acquired daily to evaluate the uniformity of the system. Spatial resolution and linearity should be evaluated weekly with a bar phantom or other suitable resolution tool. At less frequent intervals, collimator integrity, detector sensitivity, and multiple window spatial registration should be checked (*3–6*).

The QC tests specific to SPECT imaging include a high count uniformity flood, a center of rotation (COR) evaluation, three-dimensional uniformity and resolution, pixel size, and matrix linearity evaluation.

The uniformity requirements for SPECT are more stringent than those for planar imaging. Even a small amount of nonuniformity in the system will cause circular or bull's-eye ring artifacts in reconstructed images. A high count flood is used to field correct the acquisition data before reconstruction. This flood contains approximately 10,000 counts per pixel. The flood for a 64×64 matrix study should have 30–40 million counts and 120–160 million counts for a matrix size of 128×128. Depending on the stability of the system, the high count flood may be acquired daily, weekly, or biweekly (*4,7*).

The COR is the central point around which the camera head(s) rotates. The COR test measures the camera's ability to see this point and place it in the center of the

TABLE 8.1

National Electrical Manufacturers Association Performance Measurements for Scintillation Cameras

Intrinsic spatial resolution
Intrinsic spatial linearity
Intrinsic energy resolution
Intrinsic flood field uniformity
Intrinsic count rate performance
Multiple window spatial registration
System spatial resolution with and without scatter
System planar sensitivity
System volume sensitivity
System count rate performance
Reconstructed spatial resolution with and without scatter
Reconstructed image uniformity
Detector shielding

image matrix in each projection through 360° of rotation. The camera placement of the COR should be within ±0.5 pixels of the center of the matrix. COR misalignment may be the result of mechanical or electronic changes that occur suddenly or over time. A deviation in COR placement will cause blurring of the reconstructed images. The COR test should be performed at least once a week, preferably daily (*4,7*).

Imaging and reconstructing a uniform volume of activity and a three-dimensional resolution phantom on a periodic basis such as quarterly or semiannually provides information on the complete SPECT imaging system, including both hardware and software. This test is especially useful in providing information on the integrity or degradation of the system over a long period of time (*6,8*).

Pixel size can be checked periodically by imaging something of known size and then determining how many pixels are occupied by its space in the image (e.g., 10 cm/25 pixels = 0.4 cm/pixel). Pixel size measurements should be made to verify that the matrix is linear. A linear matrix has a consistent pixel size throughout the field of view. For a given matrix the pixel size should be checked in the center and all four quadrants of the field of view and in both the x and y directions. Errors in the pixel size and/or matrix linearity will affect image quality. Although error is more likely to occur in an older system, this QC test should be performed at some regular interval (such as semiannually) on all systems (*6–8*).

All QC data should be carefully reviewed at the time it is obtained, compared with previous similar data, and documented for future reference. Questions and problems should be reported promptly to the manufacturer and corrected with whatever service required. A good QC program will include but not be limited to the tests described previously.

Planar Imaging

Planar imaging alone is not used routinely to evaluate coronary artery disease. It is often used as an adjunct to SPECT imaging to help identify attenuation artifacts and determine a thallium heart–lung ratio.

When planar imaging is used exclusively, a minimum of three views should be acquired. They are the anterior, a best septal left anterior oblique (LAO; approximately 45°), and a 70° LAO or left lateral view. The patient's position for the anterior and LAO views is supine with the left arm extended over the head. The left lateral view is best acquired with the patient in the right lateral decubitus position, which minimizes breast and subdiaphragmatic attenuation. Care should be taken to reproduce comparable rest and stress views and avoid positioning variations and artifacts. Exactly the same patient positions and camera angles should be used for both the stress and rest images. Female patients should be imaged without bras for both sets of images. For all images, the camera should be as close to the chest as possible without exerting any pressure.

A minimum of 500,000 counts should be acquired for each image. The majority of counts should come from the heart, which should occupy the center of the field of view in each image. A low-energy, all-purpose or high-resolution, parallel-hole collimator can be used (*4,9*). A gated acquisition of each view can also be acquired using eight or 16 frames with standard multiple gated acquisition software (chapter 2).

Before planar or SPECT image acquisition, the patient's height and weight should be noted along with the bra cup size of female patients. The physician who reads the study should be given this data. This information will help him or her evaluate the potential for attenuation artifacts.

SPECT Imaging

The patient is positioned supine on the imaging table with the left arm out of the field of view. Supine imaging is routinely used and recommended. Because SPECT imaging will take 20–35 minutes, it is important to help the patient maintain his or her position as comfortably as possible to minimize motion artifacts. This can be accomplished by having the patient raise both arms up, fold them behind the head or over the forehead, and/or using velcro straps for support. The importance of holding still and the length of the acquisition should be explained to the patient.

SPECT acquisition also can be accomplished with the patient in the prone position, which is more comfortable for some patients. The prone position produces a greater distance between the heart and the detector and results in count reduction.

Both stress and rest acquisitions should be acquired in the same position, whether supine or prone. Again, as with planar imaging, care should be taken to reproduce comparable rest and stress studies. Female patients should be imaged without bras for both sets of studies.

An acquisition of 180° is preferable to one of 360°, because the heart lies anteriorly in the chest, with a significant amount of attenuation posteriorly. A low-energy, all-purpose collimator or high-resolution collimator can be used. A high-resolution collimator is recommended for ^{99m}Tc agents (*4,10*).

The detector should be as close to the chest as possible for all views. The patient should be centered in the field of view to minimize detector distance at all angles. A circular or elliptical orbit can be used. An elliptical orbit is better contoured for the body, reducing patient–detector distance. Care should be taken to ensure that each detector head is leveled before acquisition begins. If a detector head is not properly leveled, the reconstruction will be blurred.

A minimum of 32 views should be acquired for 180° acquisition. The use of 64 views does not improve image quality over 32 views if the time per frame is shortened to accommodate the increased number of views. The time per view (20–40 sec/stop) and subsequent total acquisition time (20–35 min) will vary, depending on the number of detector heads, radiopharmaceutical used, and dose administered. A thallium acquisition should be longer than a ^{99m}Tc acquisition. A low-dose ^{99m}Tc acquisition will be longer than a high-dose ^{99m}Tc acquisition. The maximum total acquisition time is determined by what a patient can reasonably tolerate. Although longer acquisition times yield more counts, the study may be flawed by patient movement as a result of discomfort.

Multihead detectors will provide improved resolution over a single-head detector through an increase in statistics. For a 180°-orbit cardiac SPECT, a triple-head camera improves sensitivity by 50%. Two heads positioned at 90° to each other double the sensitivity of a single-head machine. The double-head camera with heads that can be positioned at a 90° angle to each other is the optimal equipment for 180° cardiac SPECT acquisition (*11*).

Each camera manufacturer can provide recommendations on acquisition time and matrix size (64 × 64 or 128 × 128) to obtain adequate counts/view and optimal resolution for a given dose. A technologist can modify the acquisition parameters as needed (e.g., when doses are different from those recommended by the manufacturer). Once an optimal acquisition protocol for a given study and dose has been defined, it should be routinely used and modifications of the protocol will be needed only infrequently. Extremely large patients or those who have extreme difficulty lying still may require protocol modifications. These situations should be evaluated to determine whether planar imaging would be more appropriate.

Gated SPECT Imaging

A gated SPECT acquisition can be acquired for any study, including stress or rest perfusion distributions of thallium and ^{99m}Tc agents. Higher count densities, such as with a higher dose ^{99m}Tc agent, will produce a better study. Gated SPECT imaging does not increase SPECT acquisition time. It is acquired simultaneously with routine SPECT acquisition.

A gated SPECT acquisition requires a three-lead electrocardiograph interfaced with the computer. The basic principles of gated acquisition are discussed in chapter 2. Eight to 16 frames of the cardiac cycle are generally acquired for perfusion studies. Bad beats can be rejected as with standard multiple gated acquisition.

Gated SPECT images allow for the assessment of regional and global wall motion and systolic thickening. Left ventricular volume and ejection fraction also can be derived from a gated SPECT study (*12*).

References

1. Benedetta AR. Acceptance testing and performance evaluation. In: Simmons GH, ed. *The Scintillation Camera.* New York, NY: Society of Nuclear Medicine; 1988:110–127.
2. National Electrical Manufacturers Association. *Performance Measurements of Scintillation Cameras.* Washington, D.C.; National Electrical Manufacturers Association; 1994.
3. Kirchner PT, Siegel BA, eds. *Nuclear Medicine: Self-Study Program II Instrumentation.* Reston, VA: Society of Nuclear Medicine; 1996:21–22.
4. Garcia EV. Imaging guidelines for nuclear cardiology procedures. Part 1. *J Nucl Cardiol.* 1996;3:G5–G46.
5. Bernier DR, Christian PE, Langan JK. *Nuclear Medicine Technology and Techniques.* 4th ed. St. Louis, MO: Mosby; 1997:79–97.
6. Hines H, Kayayan R, Colsher J, et al. National Electrical Manufacturers Association recommendations for implementing SPECT instrument quality control. *J Nucl Med.* 2000;41:383–391.
7. English RJ. *SPECT Single-Photon Emission Computed Tomography, a Primer.* 3rd ed. Reston, VA: The Society of Nuclear Medicine; 1995:35–47.
8. Graham SL. The AAPM/RSNA Physics Tutorial for Residents: quality control for SPECT systems. *RadioGraphics.* 1995;15:1471–1481.
9. Strauss HW, Miller DD, Wittry MD, et al. Procedure guidelines for myocardial perfusion imaging. *J Nucl Med.* 1998;39:918–923.
10. Garcia EV, Cooke CD, Van Train KF, et al. Technical aspects of myocardial SPECT imaging with technetium-99m sestamibi. *Am J Cardiol.* 1990;66:23E–31E.
11. Galt JR. New instrumentation for cardiovascular nuclear medicine. *J Nucl Med.* 1994;35:20N–22N.
12. Mansoor MR, Heller GV. Gated SPECT imaging. *Semin Nucl Med.* 1999;29:271–278.

Chapter 9

Image Review and Orientation

This chapter reviews standard orientations and views with planar and SPECT imaging, including those with gated SPECT.

Planar Images

The heart sits in the chest at an oblique angle. The gross anatomy of the heart is described in chapter 1. The walls of the left ventricle, which are seen in the anterior, left anterior oblique (LAO; best septal view) and left lateral (or approximately 70° LAO) views are diagrammed in Figure 9.1. Figure 9.2 is a planar thallium perfusion study with normal distribution. Figure 9.3 is a normal ^{99m}Tc-sestamibi planar perfusion study.

SPECT Images

Figure 9.4 shows the standard orientation of sectional views of the body made from tomographic imaging. A transverse plane cuts horizontally and divides an organ from inferior to superior. A sagittal plane is a longitudinal section that divides an organ from right to left. The coronal plane is also longitudinal and divides from front to back.

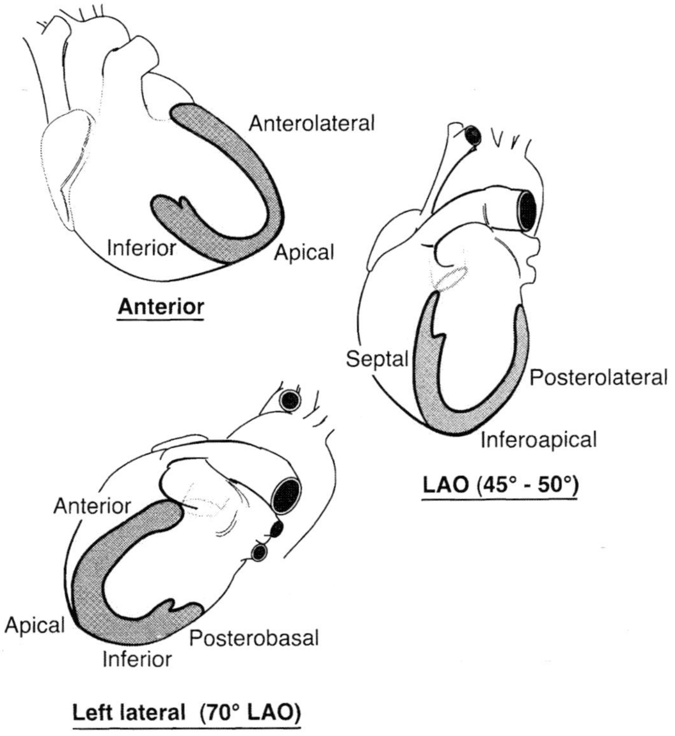

Figure 9.1 Diagram of planar views of the heart.

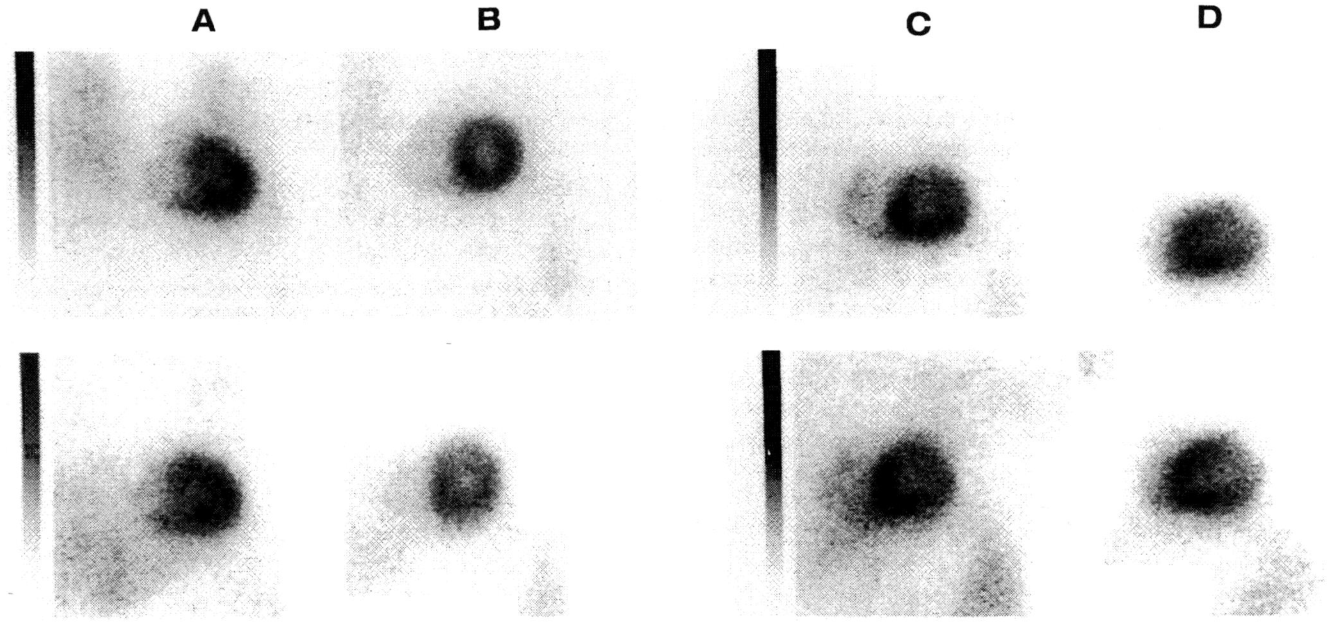

Figure 9.2 Normal thallium planar perfusion study. Top: planar thallium stress images. Bottom: the planar thallium redistribution/rest images. (A) Anterior. (B) 30° left anterior oblique (LAO). (C) 60° LAO. (D) Left lateral with patient lying supine.

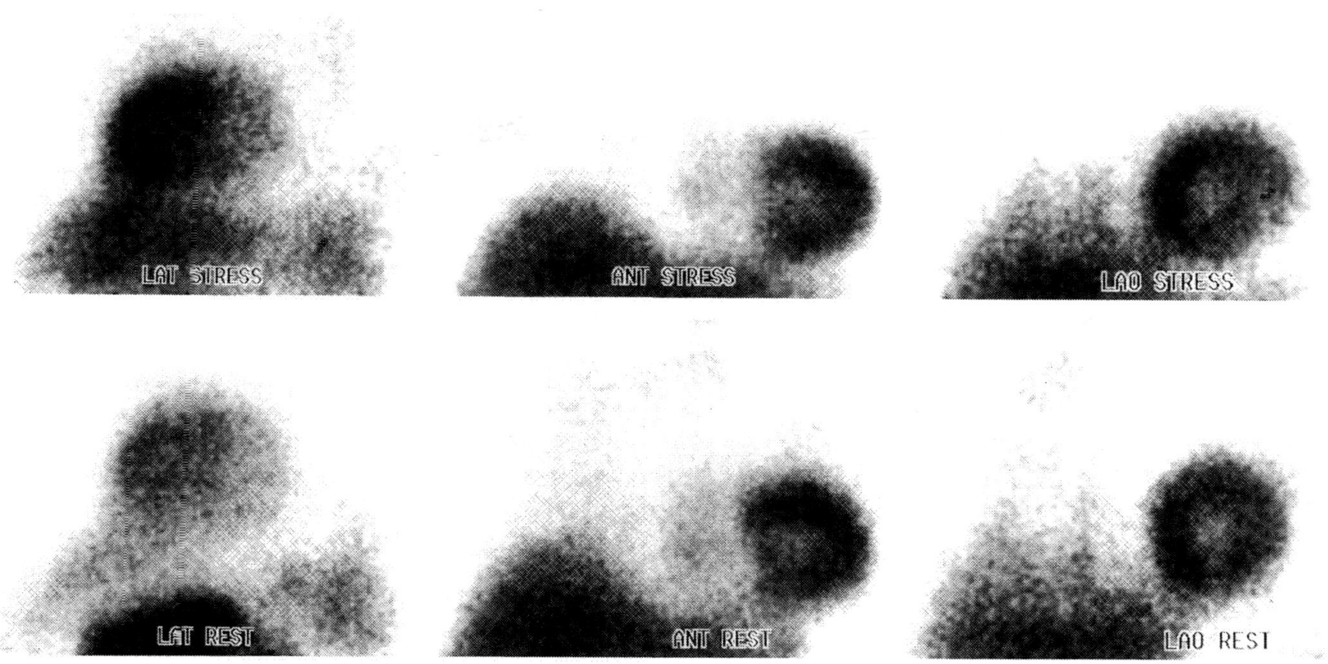

Figure 9.3 Normal ^{99m}Tc-sestamibi planar perfusion study. Top: images made after the stress injection. Bottom: images made after injection of sestamibi at rest.

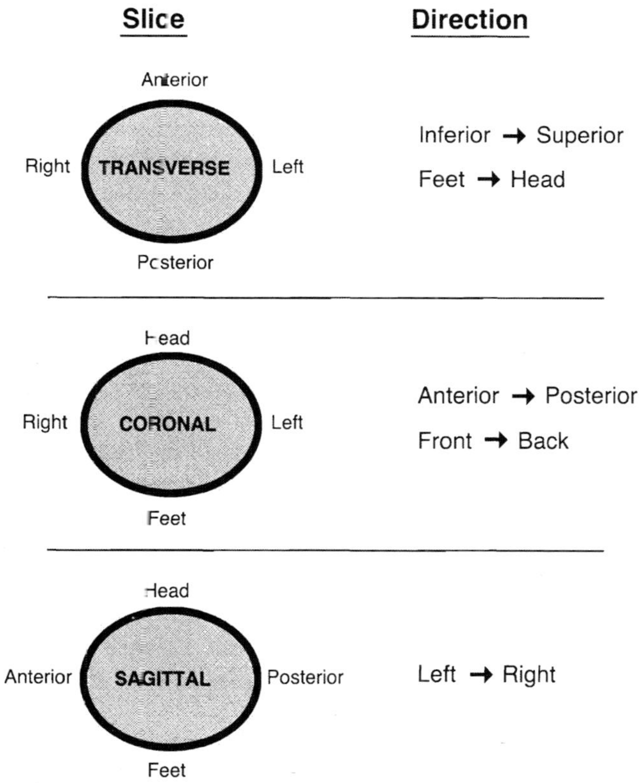

Figure 9.4 Standard orientation of sectional views of the body.

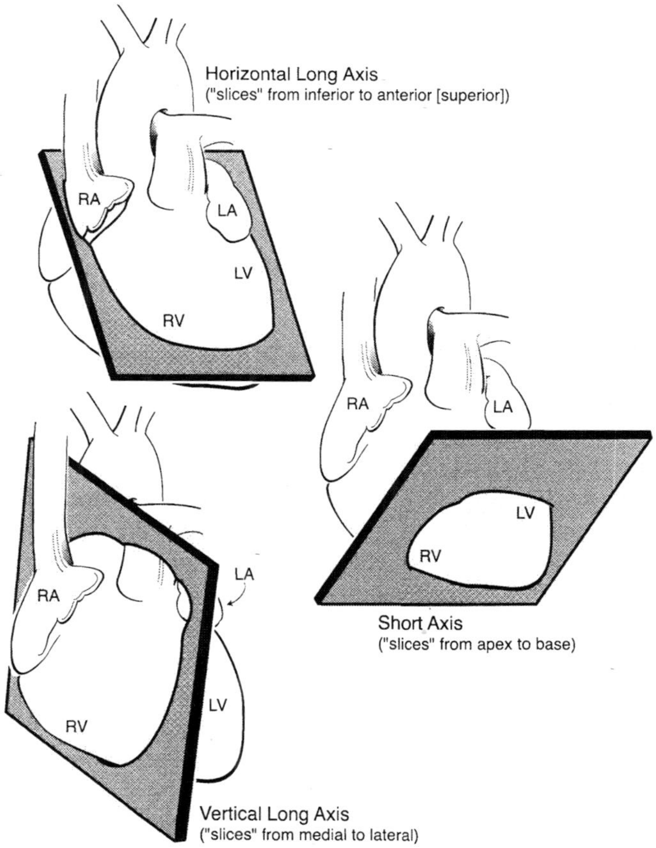

Figure 9.5 Tomographic planes of the heart.

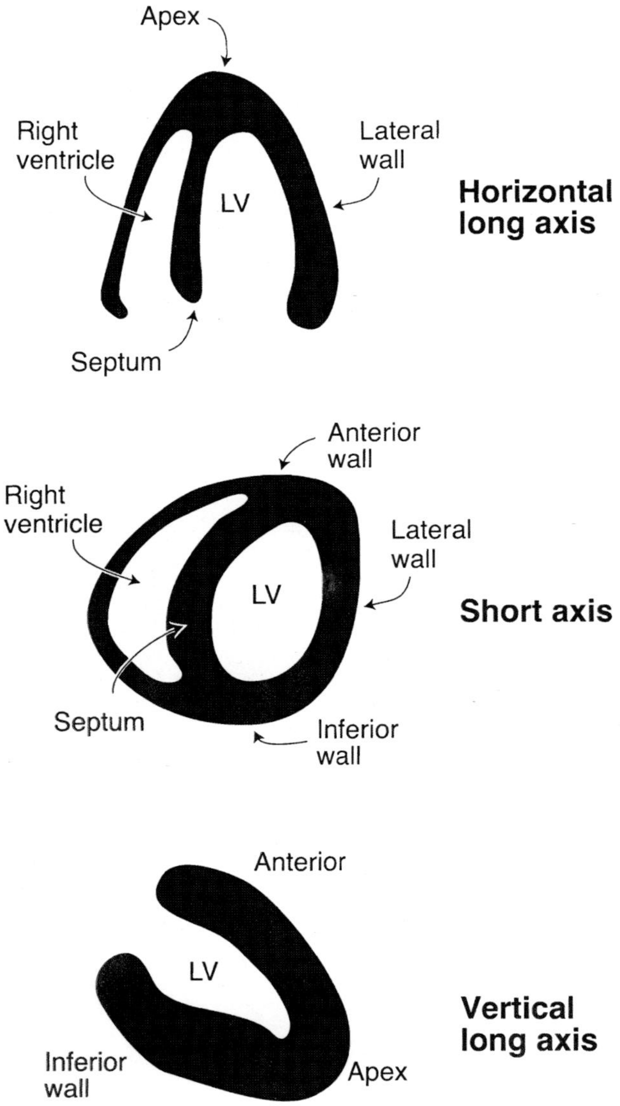

Figure 9.6 Walls of the heart seen in each sectional view.

Because of the shape of the heart and its position in the chest, sectional views of the heart are made and labeled differently from all other SPECT reconstruction. The planes along which sectional views of the heart are made are the short axis, horizontal long axis, and vertical long axis (Fig. 9.5). Short axis slices are made from apex to base and are similar to a transverse cut of the heart. Horizontal long axis slices are made parallel to the inferior surface of the heart, from the inferior to superior (anterior). Vertical long axis slices divide the heart longitudinally right to left and are made from the septum to the lateral wall. The walls of the heart seen in each of these three views are diagrammed in Figure 9.6. Figure 9.7 is a normal stress–rest SPECT perfusion study. Figures 9.8 and 9.9 are stress–rest studies showing perfusion defects. It is useful to review the images in grayscale as well as in different color scales.

By using a color scale with only five or 10 different colors, a semiquantitative analysis of the difference between stress and rest images can be made. When using only five or 10 different colors, the counts in the images are divided into 20% or 10% increments, respectively. The color of each wall at rest and stress can be compared, and the differences can be expressed semiquantitatively. Figure 9.10 shows a normal stress–rest study displayed in a five-color, 20%-increment scale and a similar study showing a perfusion defect in the same color scale.

Gated SPECT

In a gated study, the sectional views can be displayed in cine to show systolic thickening and wall motion. Figure 9.11 shows matching sectional views at end systole and end diastole. Systolic thickening and wall motion are useful parameters for evaluating functional status and myocardial viability of the heart.

Volume Images

The three-dimensional reconstructed heart (three-dimensional volume image) can also be viewed on the computer screen from any angle. The surface of the heart in systole and diastole can be viewed separately or simultaneously (Fig. 9.12). These images can be rotated in any direction to get a better view of all aspects of the heart (Fig. 9.13). If a gated study was acquired, the three-dimensional volume image also can be displayed in cine mode to assess wall motion.

Quantitative Polar Tomograms

Quantitative analysis of SPECT data is another approach to detect and/or verify perfusion abnormalities in the myocardium. These quantitative computer methods were developed largely by Cedars Sinai Medical Center (Los Angeles, CA) and Emory University (Atlanta, GA). Quantitation employs the use of a polar map or bull's-eye plot of the left ventricle, constructed by slicing the heart from apex to base and then displaying all slices together in a concentric ring manner. This display of the short axis slices provides a way to visualize the entire left ventricle in one image.

Figure 9.14 illustrates which walls of the heart are seen in each area of a polar map or plot. The center of the polar plot is the apical-most portion of the heart. The periphery of the plot is the base of the heart. The lateral wall is from the center to the right. The septal wall is from the center to the left. Superior to the center is the anterior wall, and inferior to the center is the inferior wall. Figure 9.15 shows the territories in the plot supplied by each of the major coronary arteries.

Two polar maps are made for each patient: one for the stress study and one for the rest study. The difference between the two polar plots can be evaluated and compared with a gender-specific normal database. A normal database is constructed by averaging many polar plots from

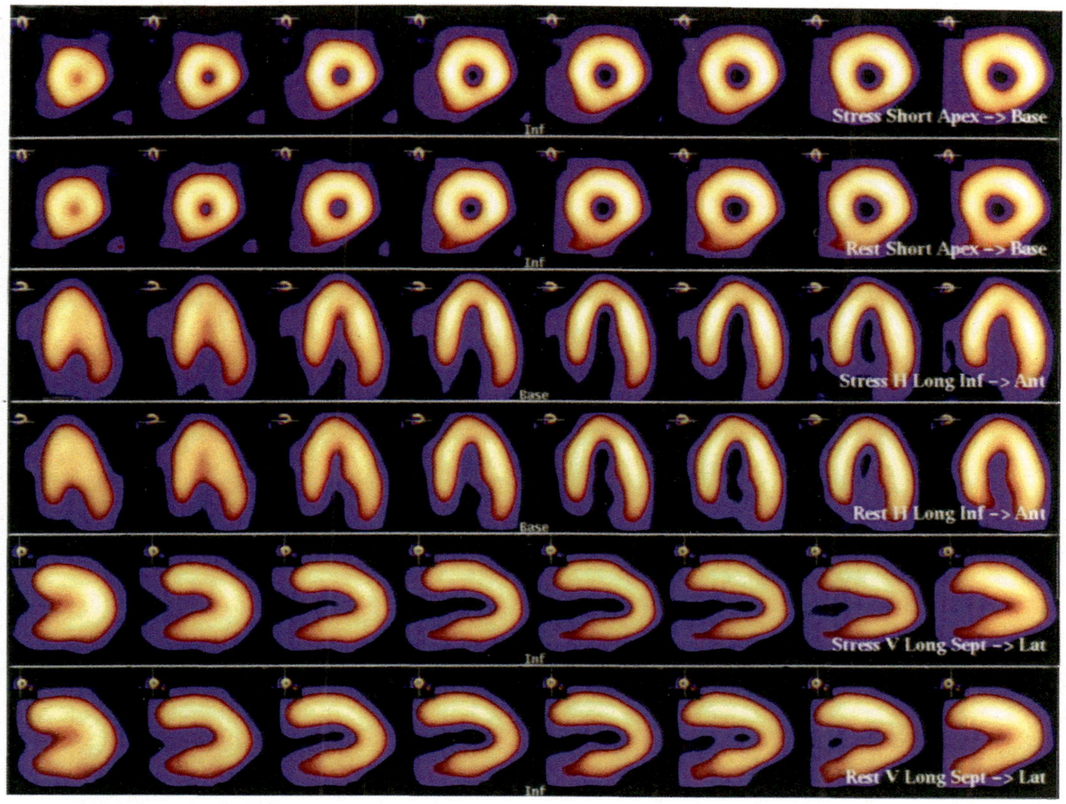

Figure 9.7 Normal stress–rest SPECT perfusion study.

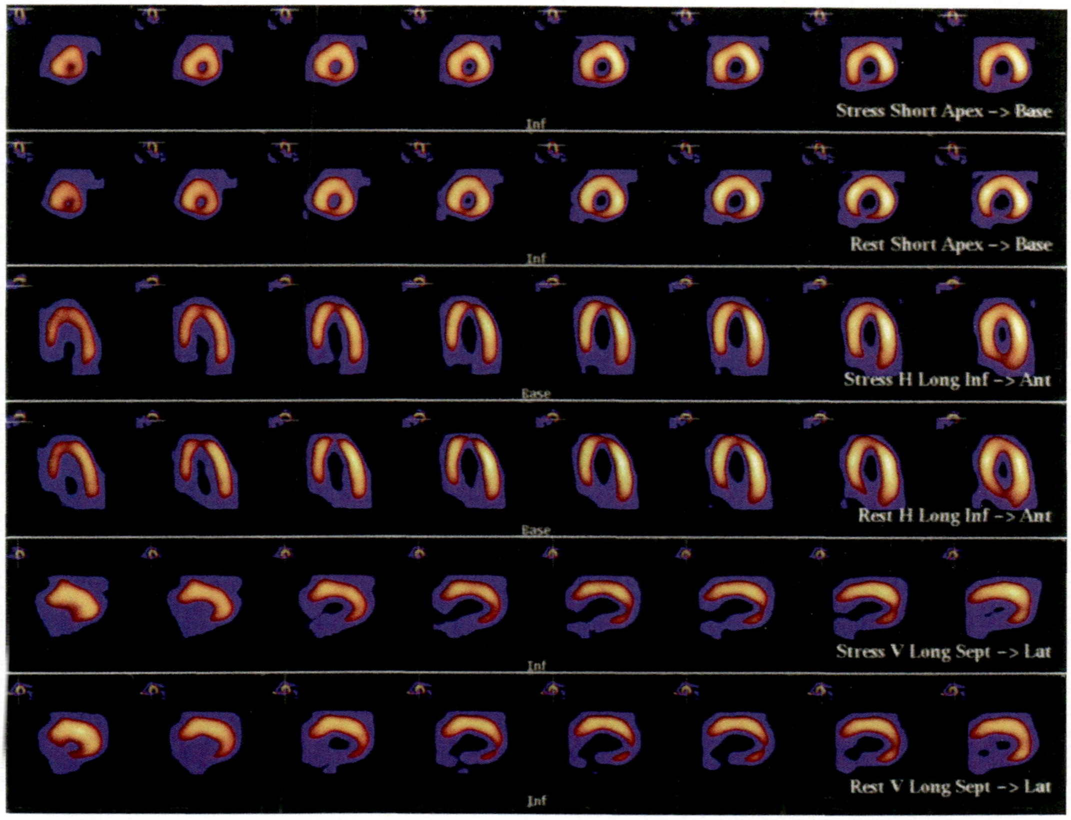

Figure 9.8 Stress–rest perfusion study with a fixed defect. A fixed inferior wall defect is seen in the short axis and vertical long axis slices. The horizontal long axis shows a small inferoapical fixed defect.

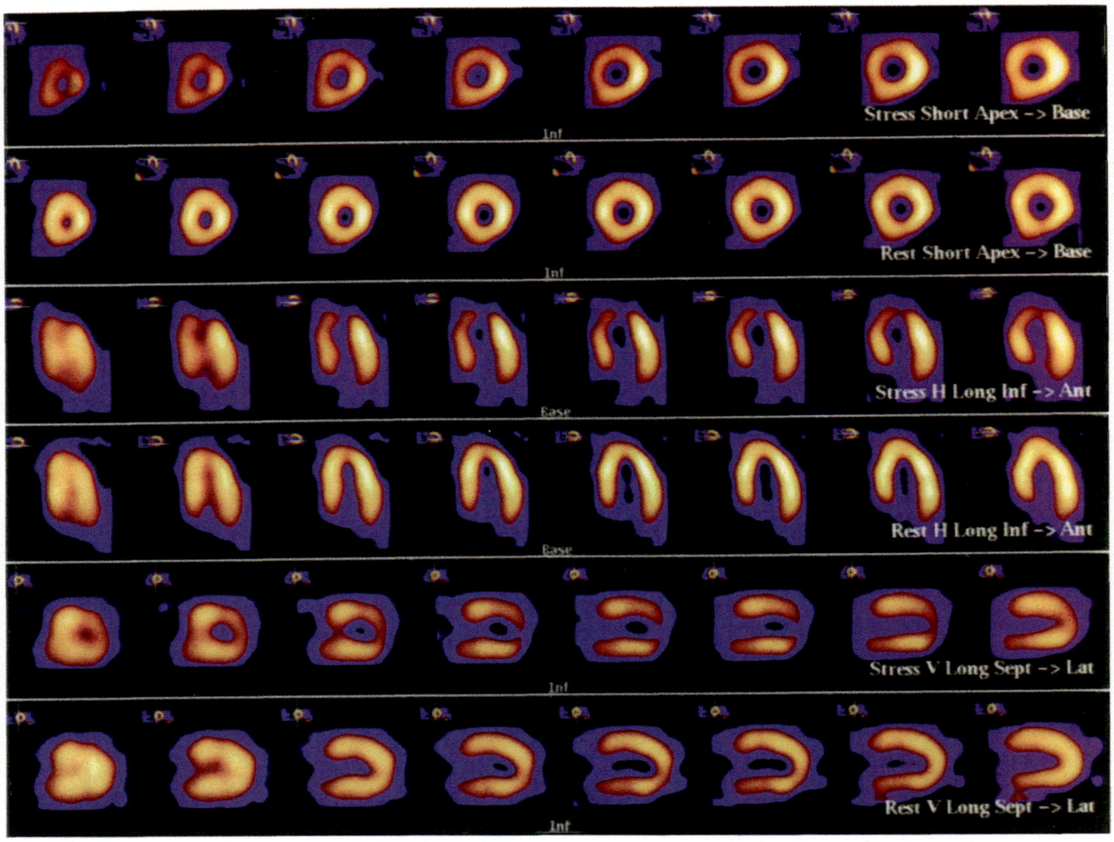

Figure 9.9
Stress–rest perfusion study with a reversible defect. The short axis slices and horizontal long axis slices show a reversible anteroseptal/septal defect. The vertical and horizontal long axis slices show a reversible anteroapical/apical defect.

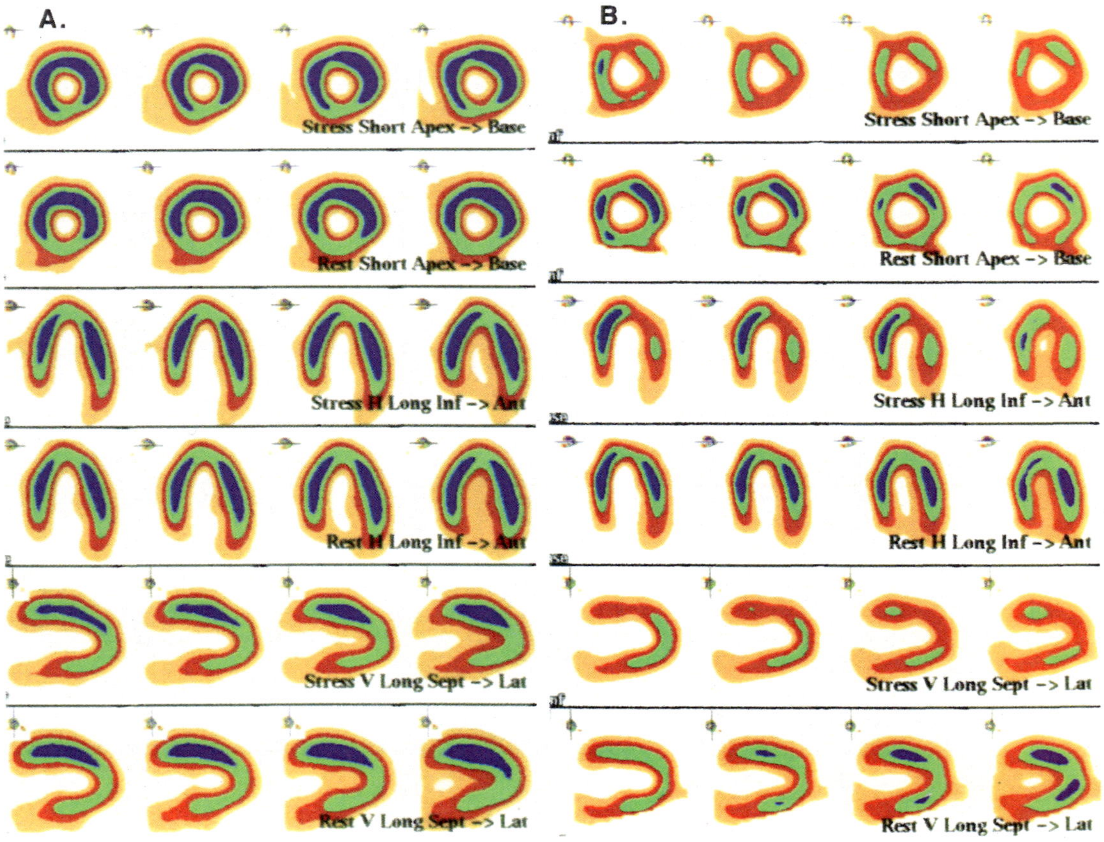

Figure 9.10
Color scale for semiquantitative display. These images are displayed in five colors on a scale of 20% increments.
(A) Normal study.
(B) Reversible defects in the inferior, lateral, and anterior walls.

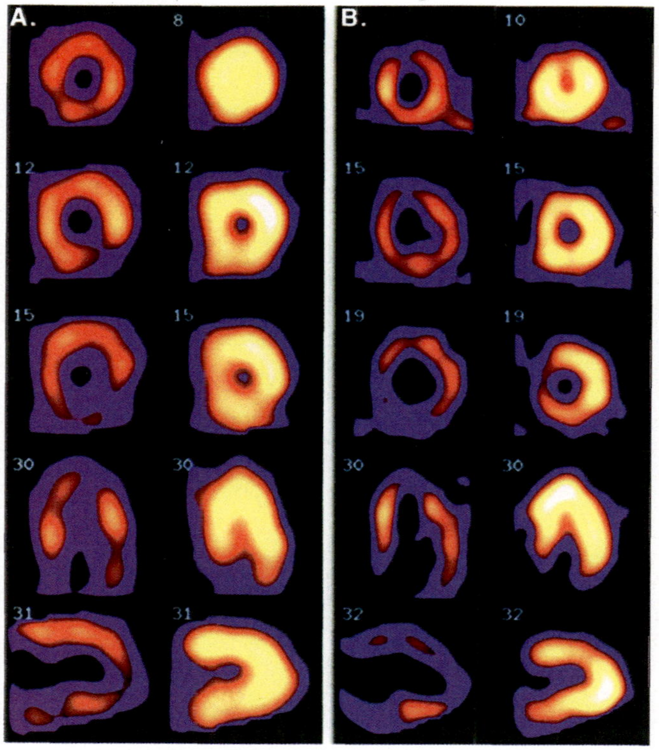

Figure 9.11 **Matching end diastolic (ED) and end systolic (ES) images from a gated SPECT study. Short axis and horizontal and vertical long axis slices in ED and ES. The ES frames depict systolic thickening of the heart walls. (A) Normal study. (B) Decreased thickening in the anteroapical segment.**

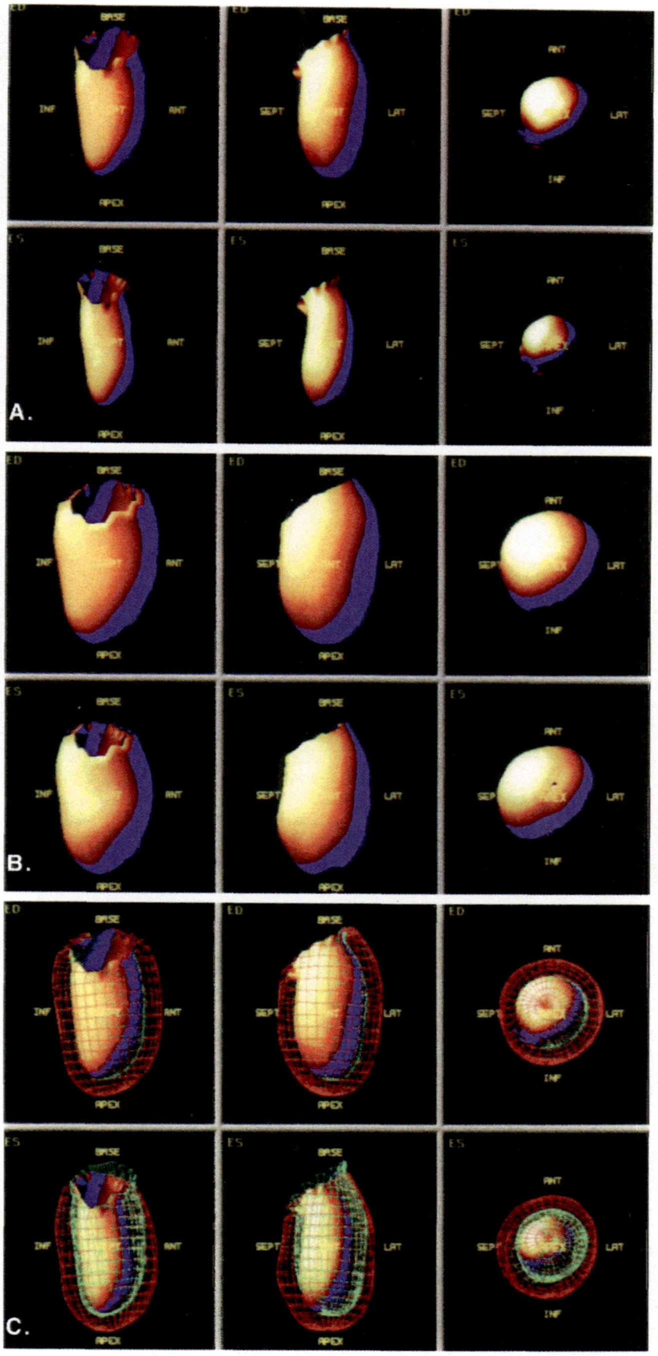

Figure 9.12 Three-dimensional volume images. (A) Systole. (B) Diastole. (C) Both.

healthy patients or patients with low likelihood of coronary artery disease.

Quantitation was developed first for thallium. The short axis of the heart is divided into 15 slices of equal thickness. With the exception of several slices at the apex, each slice is divided into 40 sectors of 9° each. The maximum counts per sector are plotted circumferentially from apex to base. Because the counts per pixel in the two studies will vary, the delayed thallium study plot must be normalized to the stress study plot so that the two can be compared (*1*).

The stress study plot and the delayed study plot are compared to generate another plot showing the extent of reversibility. The reversibility plot is generated by subtracting the normalized delayed study plot from the stress study plot. Defects that redistribute and defects that do not redistribute are shaded differently. The degree of reversibility is also displayed by the gray- or color scale (*1,2*).

The first three plots described here are referred to as the raw data plots. These are compared with the gender-specific normal database plots to generate blackened pixel plots and standard deviation plots. A blackened pixel plot shows which areas are greater than a specified standard deviation below the mean. The blackened area represents the size and shape of the perfusion deficit. The standard deviation plot color codes the myocardium based on the number of standard deviations below the mean of the normal database.

With the original polar plot construction method described previously, the apex of the heart is especially prone to artifacts. The boundary of the apex or apical slice is de-

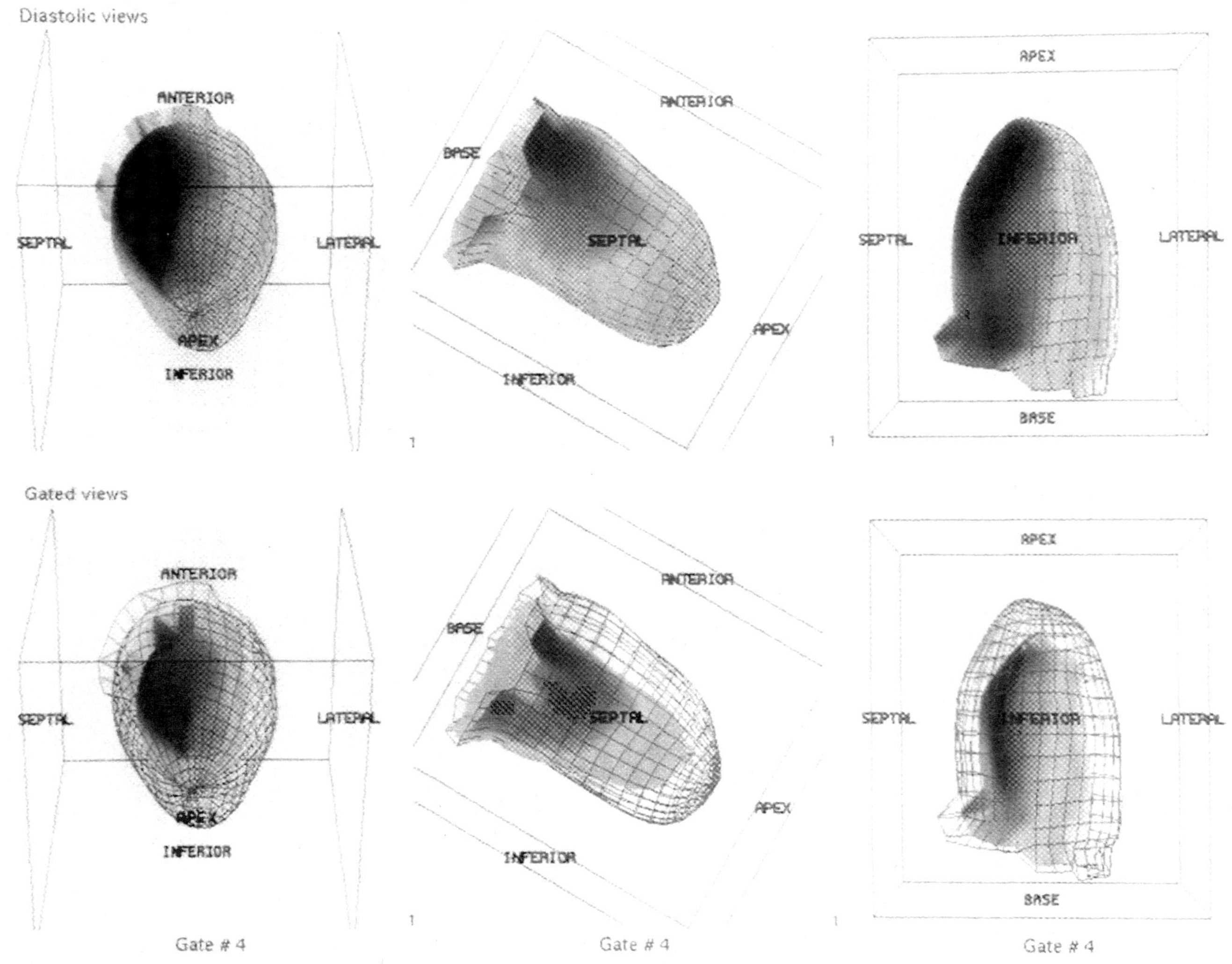

Figure 9.13 Three-dimensional volume image. Different angles of the heart.

Medical diagnostic images courtesy of Marconi Medical Systems, Inc.

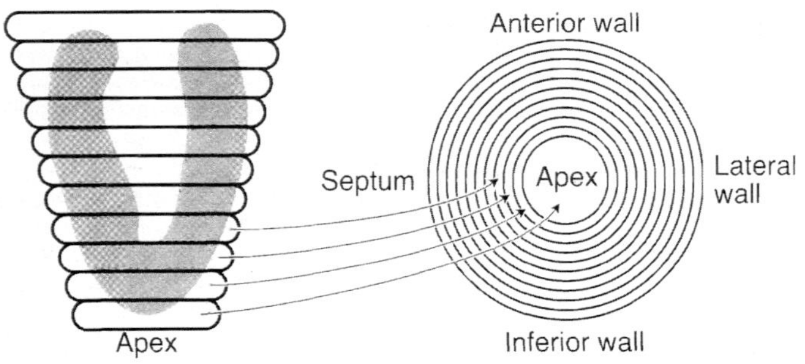

Figure 9.14 Walls of the heart seen on polar tomograms.

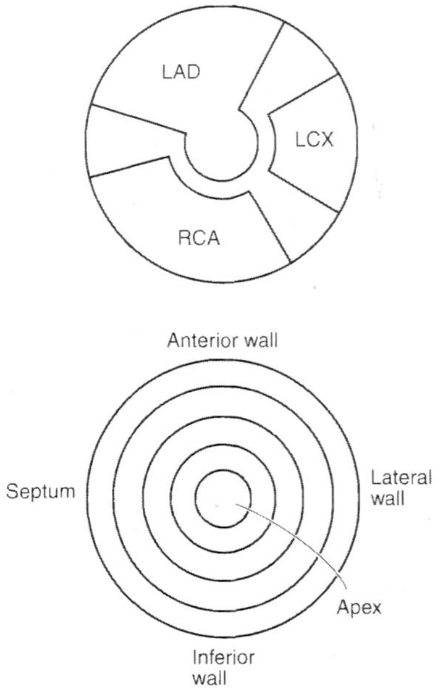

Figure 9.15 Coronary artery territories seen on polar tomograms.

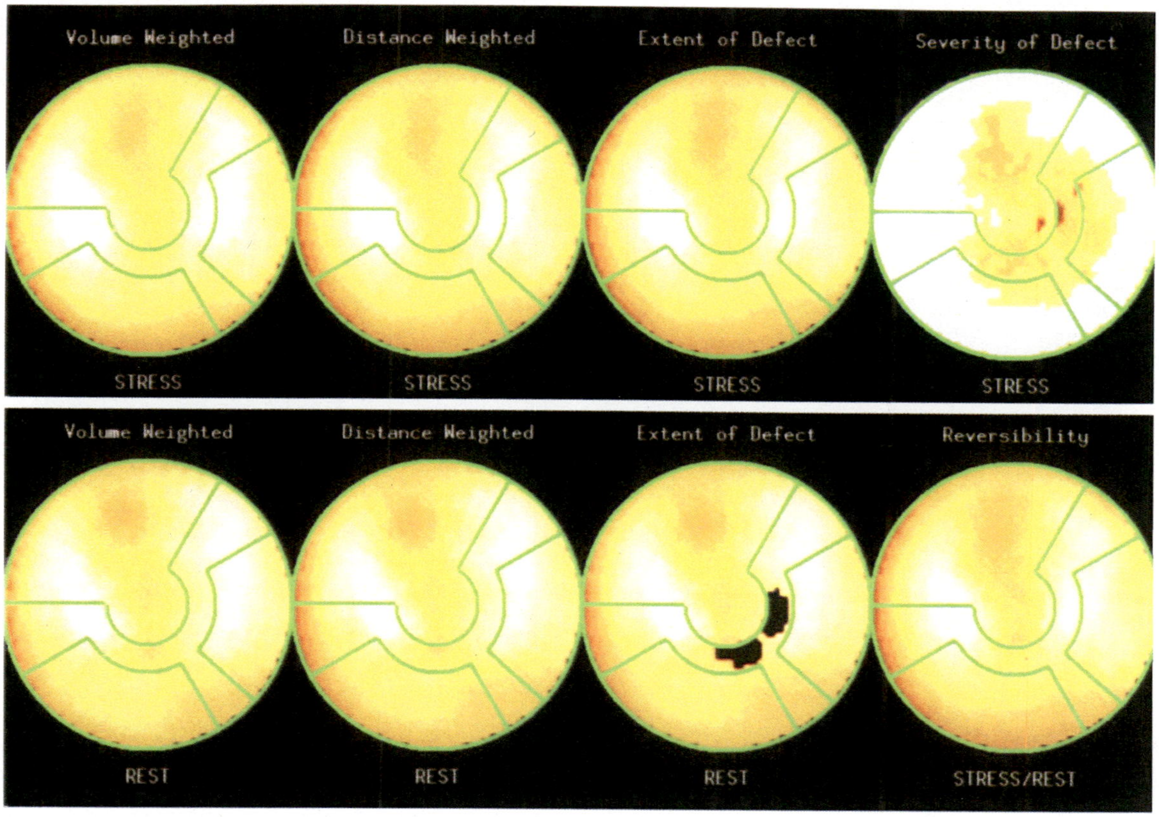

Figure 9.16 CEqual polar plot. Normal study.

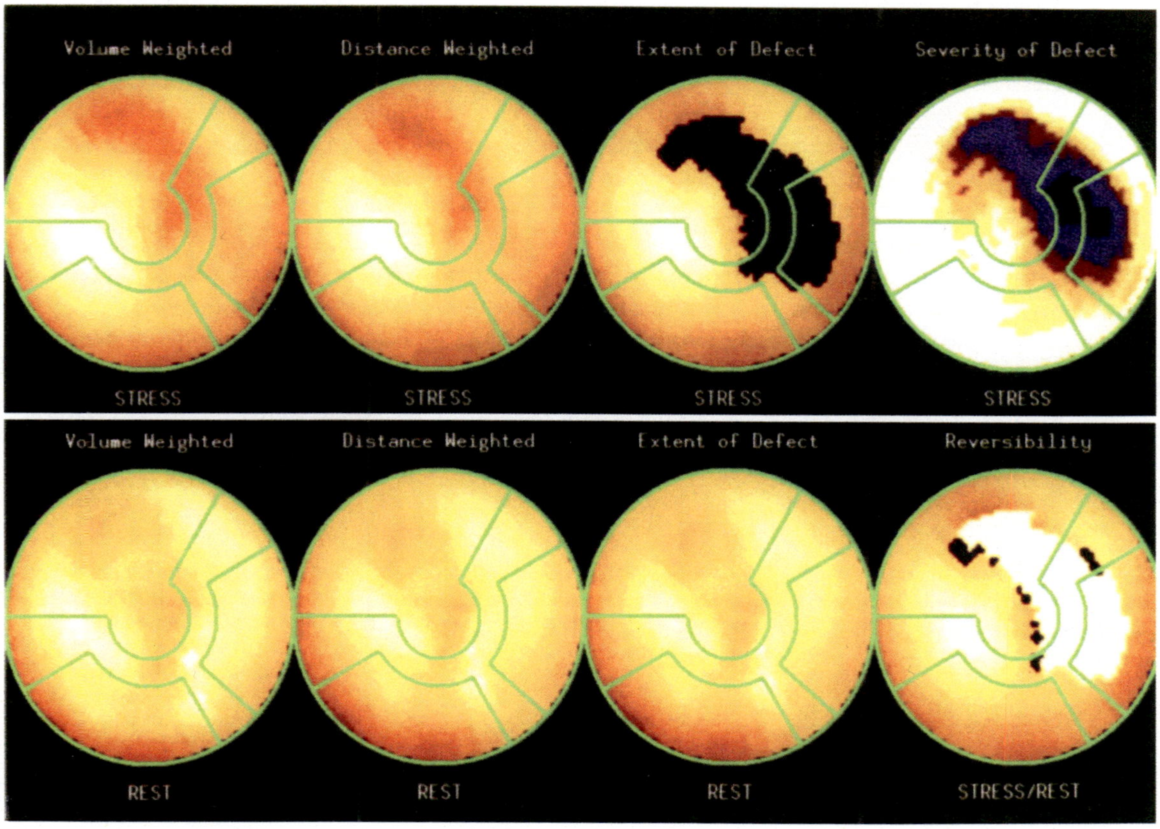

Figure 9.17 CEqual polar plot showing anterior and lateral wall ischemia.

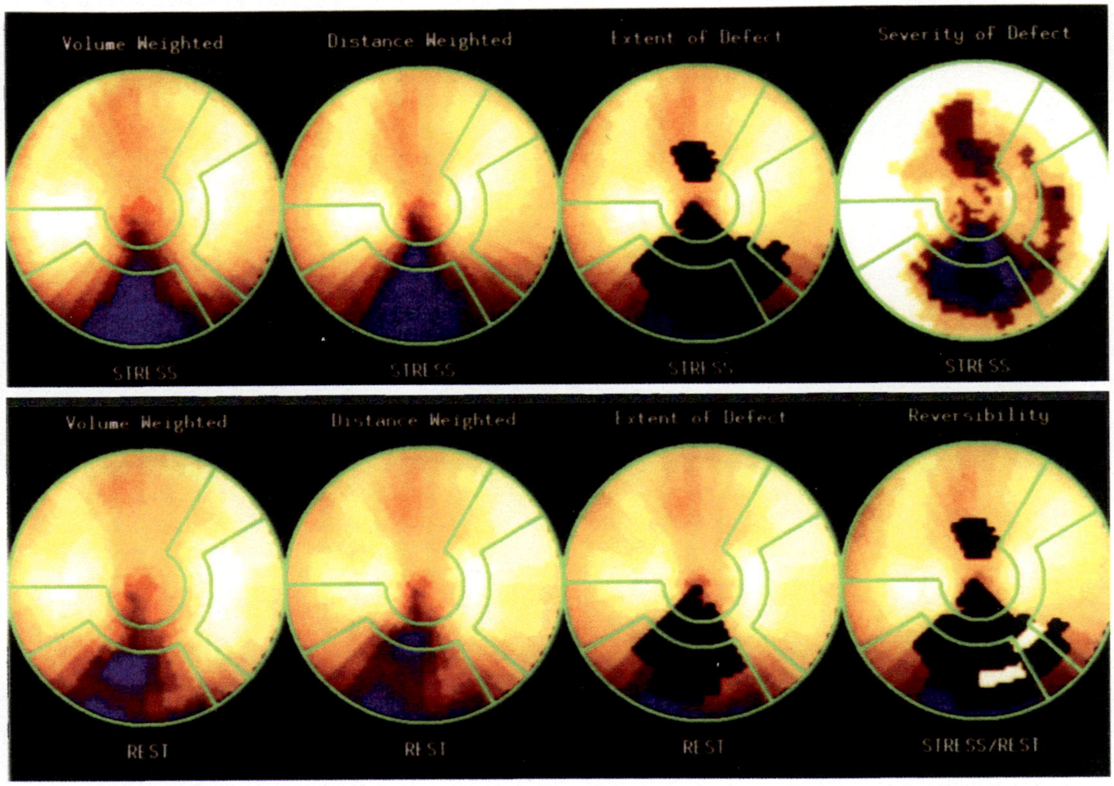

Figure 9.18 CEqual polar plot demonstrating an inferior wall fixed defect.

termined manually. If not selected properly, the apical region may include background. The apical area of the plot then will have a decreased count density.

With improvements in computer technology and the development of technetium agents came another technique to generate count profiles for quantitation and polar map displays. This newer technique uses three-dimensional data and sampling to generate circumferential profiles (*3,4*). These polar plots for stress and rest can be compared with each other and with normal databases. Gender-specific databases have been developed for both technetium and thallium. The newer method was developed by Cedars Sinai and Emory and is known as "CEqual" (for Cedars and Emory quantitative analysis).

The new maps have 12 slices or regions divided into 40 sectors each. The distance-weighted polar map is similar to the original polar plots for thallium in that every ring in the map is the same width (i.e., each patient's stress and rest study is divided into 12 slices of equal thickness). The volume-weighted polar map makes a defect area on the two-dimensional display equal to the size of the three-dimensional volume of the defect. Examples of polar plots generated by the CEqual program are seen in Figures 9.16–9.18. Other plots shown include the extent of defect, severity of defect, and reversibility.

References

1. Gabor FV, Datz FL, Christian PE, et al. Computer-assisted diagnosis of cardiac perfusion studies. *J Nucl Med Technol.* 1991;19:238–244.
2. Klein JL, Garcia EV, DePuey G, et al. Reversibility bull's-eye: a new polar bull's-eye map to quantify reversibility of stress-induced SPECT thallium-201 myocardial perfusion defects. *J Nucl Med.* 1990;31:1240–1246.
3. Garcia EV, Cooke CD, Van Train KF, et al. Technical aspects of myocardial SPECT imaging with technetium-99m sestamibi. *Am J Cardiol.* 1990;66:23E–31E.
4. Garcia EV, Van Train K, Maddahi J, et al. Quantification of rotational thallium-201 myocardial tomography. *J Nucl Med.* 1985;26:17–26.

Chapter 10

Causes of Artifacts and Resolution Techniques for SPECT Imaging

Many factors can affect the quality of a myocardial perfusion SPECT study. Artifacts can be classified as technical or patient related. Table 10.1 lists the specific artifacts covered in this text.

Technical Artifacts

The importance of camera quality control cannot be overstated. Camera quality control is discussed in chapter 8. Even with a quality control program in place, a technologist must be alert to unusual or intermittent problems that can arise at any time, even with newer equipment. Unexplained artifacts, such as a line, half moon, or other shape in the field of view, can arise from electronics or software. The system may need to be rebooted or serviced. Figure 10.1 is an example of an electronic artifact that appeared suddenly several times in a new, state-of-the-art camera that was less than 1 year old.

The artifacts caused by deficiencies in imaging system performance are also discussed in chapter 8. These include artifacts caused by flood field nonuniformity and center of rotation errors. Proper and consistent imaging parameters and procedures will minimize artifacts caused by camera set-up and acquisition. Artifacts caused by an improper data reconstruction technique, such as improper oblique angle selection or filter selection, are also discussed in chapter 8.

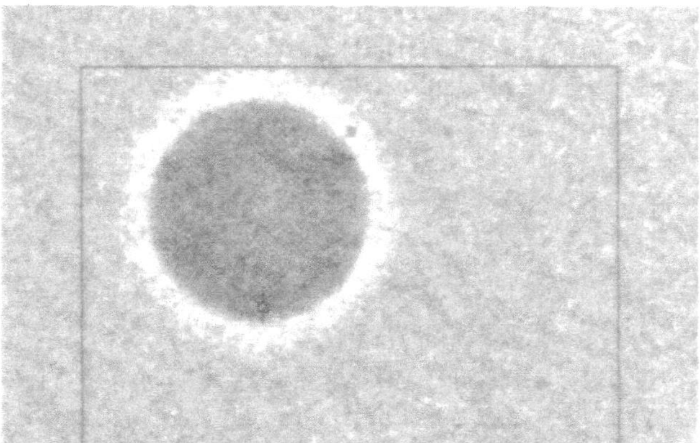

Figure 10.1 Electronic artifact in flood image.

TABLE 10.1
Causes of Common Artifacts in Myocardial Perfusion Imaging

Technical artifacts
System nonuniformity
Wrong center of rotation
Improper oblique angle selection for reconstruction
Wrong filter selection
Patient-related artifacts
Upward creep of the heart
Breast attenuation
Lateral chest wall fat attenuation
Diaphragmatic attenuation
Overlying and adjacent abdominal visceral activity
Patient motion

Patient-Related Artifacts

Upward Creep of the Heart

Upward creep of the heart may occur during an exercise stress study acquisition performed shortly after the termination of exercise. This problem is more likely to occur on a thallium stress study than on a ^{99m}Tc agent study, because the acquisition of ^{99m}Tc agent studies is usually delayed for at least 30 minutes after treadmill exercise. Upward creep of the heart is a gradual upward movement of the heart during acquisition. It is caused by a change in the depth of respiration. It is associated with exercise to an exhaustive level and is more significant in women and patients with chronic obstructive pulmonary disease. During and shortly after exercise, a patient may be breathing deeply. The transient increase in mean total lung volume after exercise causes diaphragmatic flattening, which results in a mean lower position of the diaphragm and heart. With rest, the diaphragm and heart assume the normal position in the chest.

As a result of upward creep, the position of the heart may change as little as 2 pixels or as much as 8 pixels during acquisition. Reconstruction of acquisition data with upward creep will result in false-positive perfusion defects.

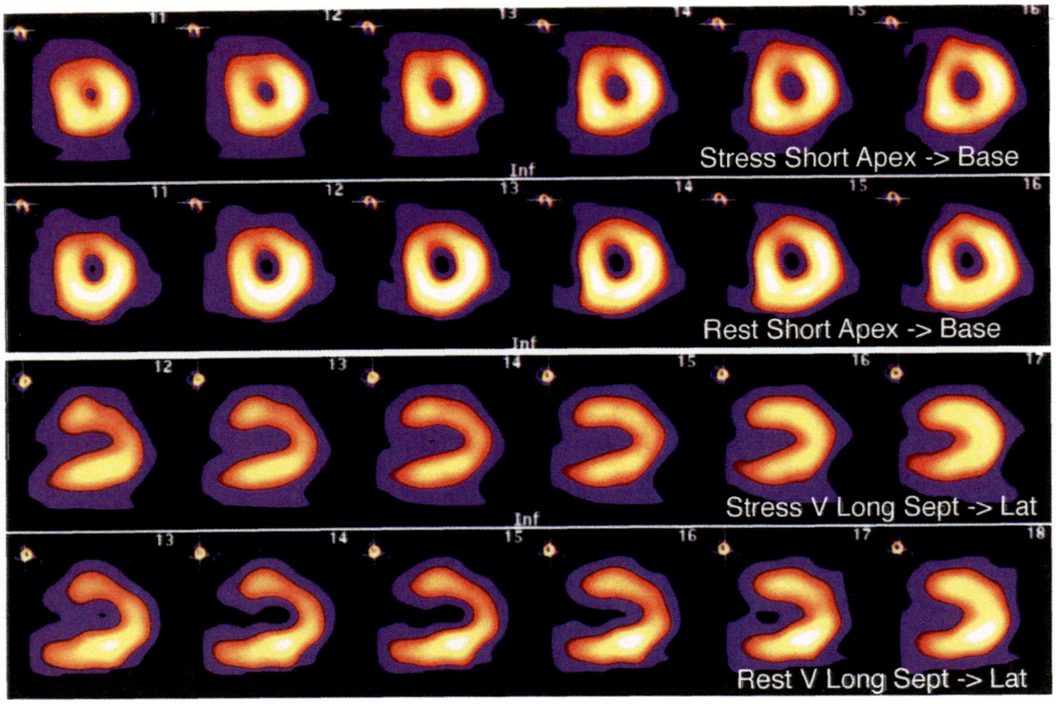

Figure 10.2 SPECT myocardial perfusion study with breast attenuation. Anterior wall attenuation seen on short axis and vertical long axis slices in both stress and rest images.

Upward creep of at least 2 pixels will produce an anteroseptal defect. Upward creep of more than 4 or 5 pixels will produce an inferoseptal defect.

To minimize the possibility of upward creep and its effects on imaging, each patient's depth and frequency of breathing should be observed and evaluated before beginning a thallium stress study acquisition. SPECT imaging occasionally may have to be delayed 5–15 minutes. If a thallium stress study is delayed for this reason, the sensitivity of the study may be affected because of the early redistribution property of thallium. A 5-minute anterior planar image can be acquired during the delay time. This planar image can be used to evaluate pulmonary thallium uptake (chapter 5, section 1) and transient ischemic dilatation of the left ventricle. Both of these parameters are useful in the overall assessment of coronary artery disease (*1,2*).

Soft Tissue Artifacts

A patient's body weight and habitus have an influence that cannot be disregarded. The technologist performing the study and the physician who will read and interpret the study must be fully aware of the patient's body type (*3*). Patient factors that should be noted are listed in Table 10.2.

Muscle and fat can cause significant attenuation of counts coming from the myocardium. Attenuation can cause a false-positive study. Often the mild-to-moderate fixed perfusion defects that appear on both the stress and rest studies are caused by attenuation. Attenuation is a problem with both thallium and ^{99m}Tc agents. The photon energy will affect the severity of the attenuation defect. Identification of attenuation artifacts improves the specificity of myocardial perfusion imaging and helps to eliminate false-positive studies.

Attenuation artifacts are more common in cardiac SPECT imaging than in any other type of SPECT imaging. Attenuation correction algorithms for a 180° orbit were not available in the past. Although newer cameras are equipped with transmission scanning and attenuation correction software packages, such packages are not widely utilized at this time. Transmission sources are expensive.

Technologists and interpreting physicians must learn to recognize attenuation artifacts and minimize their effects where possible. When the question of differentiating between an attenuation artifact and a true perfusion defect is unresolved by conventional means, a positron emission tomography (PET) study can be helpful. PET imaging uses high-energy photons, and attenuation correction is applied. PET imaging is addressed in section 4 of this text.

TABLE 10.2

Factors Influencing Count Attenuation

Sex, weight, height	Lateral chest wall fat
Chest circumference	Abdominal protuberance
Breast size	Chest wall deformity

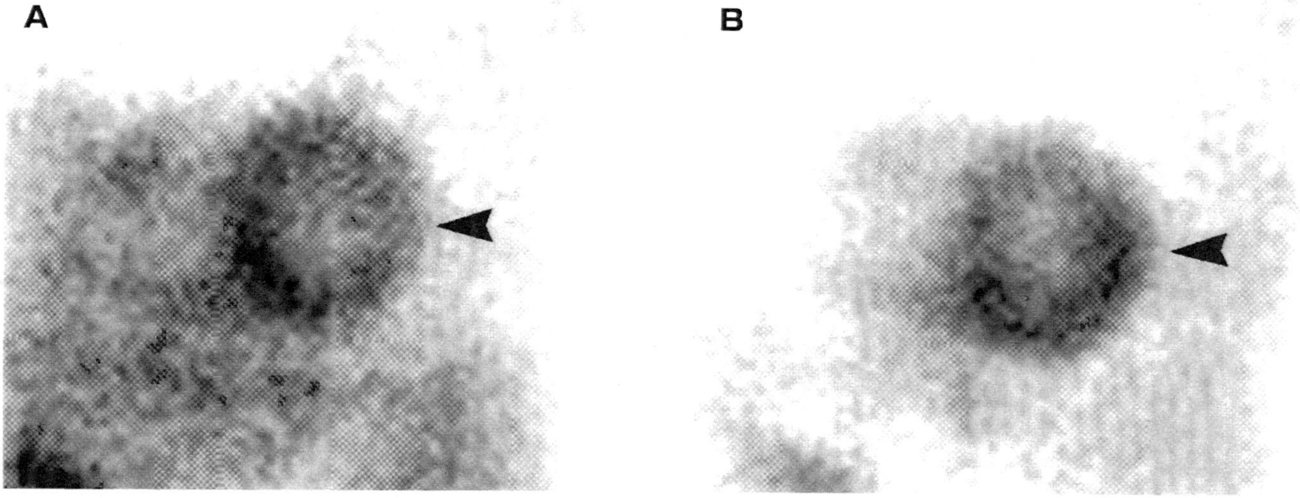

Figure 10.3 Planar views demonstrating breast attenuation. (A) Left anterior oblique (LAO) view acquired with patient supine. (B) LAO view acquired with patient in the right decubitus position. In (A), the position of the breast caused attenuation of the posterolateral wall. By changing the patient's position and the position of the breast tissue, less attenuation is seen in (B).

Breast Attenuation

In imaging women, breast tissue causes attenuation defects. The location and severity of the defect depends on the thickness, density, and position of the breast tissue. The breast is known to cause anterior, anterolateral, and anteroseptal perfusion defects. Other walls also may be affected by breast attenuation, depending upon the size and location of the breasts. The defect may appear partially reversible if the breasts are not identically positioned for stress and rest acquisitions. Rest and stress studies in the same patient should be performed consistently with the bra on or off (preferably off). If the patient is imaged with the bra on, the same bra should be worn for both studies. Figure 10.2 is a study with anterior wall breast attenuation. Breast attenuation occasionally also may be a problem with male patients.

The following steps should be taken to recognize breast attenuation artifacts:

- Record chest circumference and bra cup size;
- Observe the patient on the imaging table;
- Inspect the rotating planar images for breast shadow; and
- Observe the wall motion and thickening on the gated study.

If systolic thickening and wall motion are decreased, then the defect is more likely to be real than caused by attenuation artifact.

Another way to recognize breast attenuation is to acquire two planar anterior or oblique view images each for 400,000–500,000 counts. The patient should lie supine for the first anterior image. The second anterior image should be taken with the patient in the right decubitus position with both arms up. The wall of the heart affected by breast attenuation will look photopenic in the supine anterior view when compared with the anterior view acquired with the patient in the right decubitus position (Fig. 10.3).

Lateral Chest Wall Fat Attenuation

Lateral chest wall fat can cause a fixed cardiac lateral wall perfusion defect. The steps that should be taken to recognize this artifact and minimize its effect in the interpretation of the study are similar to those listed for breast attenuation and include:

- Record chest circumference;
- Observe patient supine on the imaging table;
- Inspect the rotating planar images; and
- Observe normal lateral wall motion and thickening on the gated study.

Diaphragmatic Attenuation

Diaphragmatic attenuation will vary from the stress to the rest study if the diaphragmatic height changes. Diaphragmatic attenuation most commonly causes a fixed inferior wall defect (Fig. 10.4) and is common in obese patients and accentuated in patients with ascites and those on peritoneal dialysis. The following steps should be taken to recognize this artifact:

- Observe abdominal protuberance;
- Inspect rotating planar images;
- Observe inferior wall motion and thickening on the gated study; and
- Obtain two planar anterior views, one in the supine position and one in the prone position.

When the patient is imaged in the prone position, the arms preferably should be crossed under the head as opposed to down at the sides. This position "lifts the heart" and causes less inferior wall attenuation. The prone image may demonstrate more anterior wall attenuation. Figure 10.5 is an example of planar images acquired to evaluate for diaphragmatic attenuation.

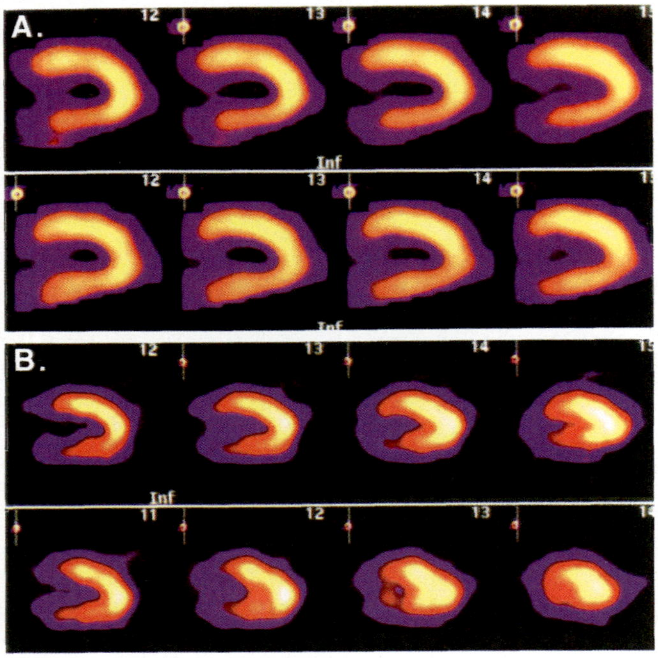

Figure 10.4 SPECT myocardial perfusion study with inferior wall attenuation. (A) shows minimal diaphragmatic attenuation of the inferior wall. (B) also shows an inferior wall fixed defect that demonstrated normal wall motion in cine. The brightness of the inferior wall is decreased compared with the anterior wall. This is caused by diaphragmatic attenuation.

Overlying and Adjacent Abdominal Visceral Activity

Overlying abdominal visceral activity can cause an apparent increase in activity in the inferior wall of the heart. The same can be caused by Compton scatter from adjacent abdominal viscera. The remainder of the heart appears to have lower count density, resulting in artifactual defects when stress and rest studies are compared (*3*).

^{99m}Tc-sestamibi activity in the viscera (liver, bowel, and gallbladder) often exceeds that of the heart (Fig. 10.6). Abdominal visceral activity is more pronounced with pharmacologic stress. In a delayed thallium study, the bowel may be problematic because of bowel excretion of thallium. Thallium is also known to concentrate in the liver and stomach. The amount of uptake in visceral organs will vary from patient to patient and study to study.

Several tasks should be performed routinely to avoid artifactual defects from overlying and/or adjacent abdominal visceral activity:

- When positioning the patient before beginning the SPECT acquisition, the technologist must observe for and recognize increased activity in the liver, bowel, or other visceral organs. Milk, water, or a fatty meal should be given to the patient to displace bowel activity before starting SPECT acquisition. These may also be administered to promote hepatobiliary clearance;
- Before starting the SPECT acquisition, visceral organs can be masked with a lead apron or shield. Caution must be taken not to mask any of the inferior or apical wall of the heart;
- After a study has been acquired, the rotating planar images should be inspected for extracardiac activity in adjacent viscera; and
- Planar prone imaging also may be helpful.

Patient Motion

Patient motion (>0.5 pixels) during acquisition will cause a misalignment of the opposing walls of the heart on the reconstructed images (Figs. 10.7 and 10.8). The location of the artifact will depend on direction and time of motion during acquisition. The artifact is worse if the motion occurs during the middle of acquisition as opposed to at the beginning or end of acquisition. The following steps should be taken to avoid or minimize patient motion:

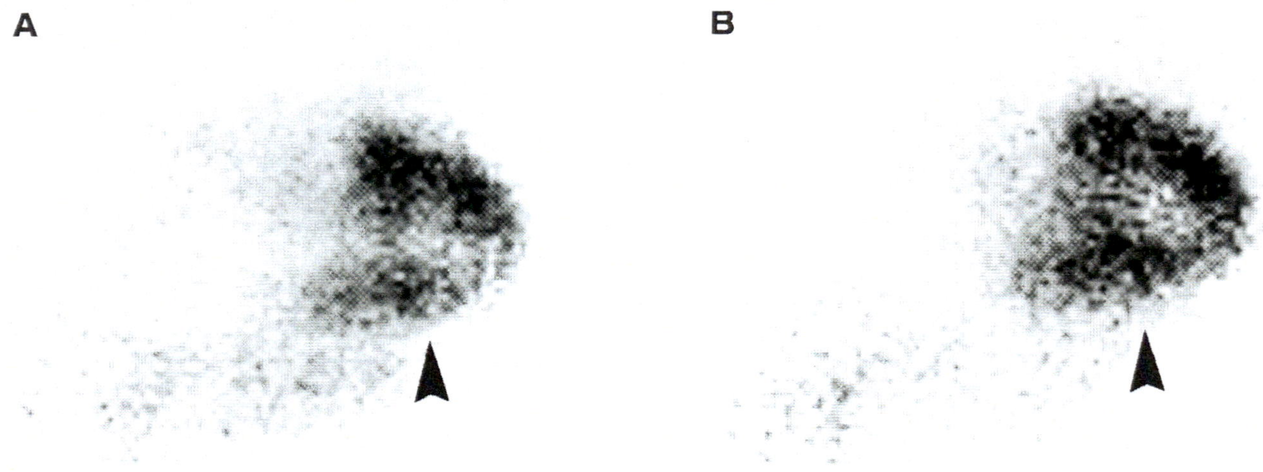

Figure 10.5 Planar views demonstrating diaphragmatic attenuation. (A) Anterior view acquired with patient in the supine position. (B) Anterior view with patient lying prone and camera under the table. View (B) shows less inferior heart wall attenuation.

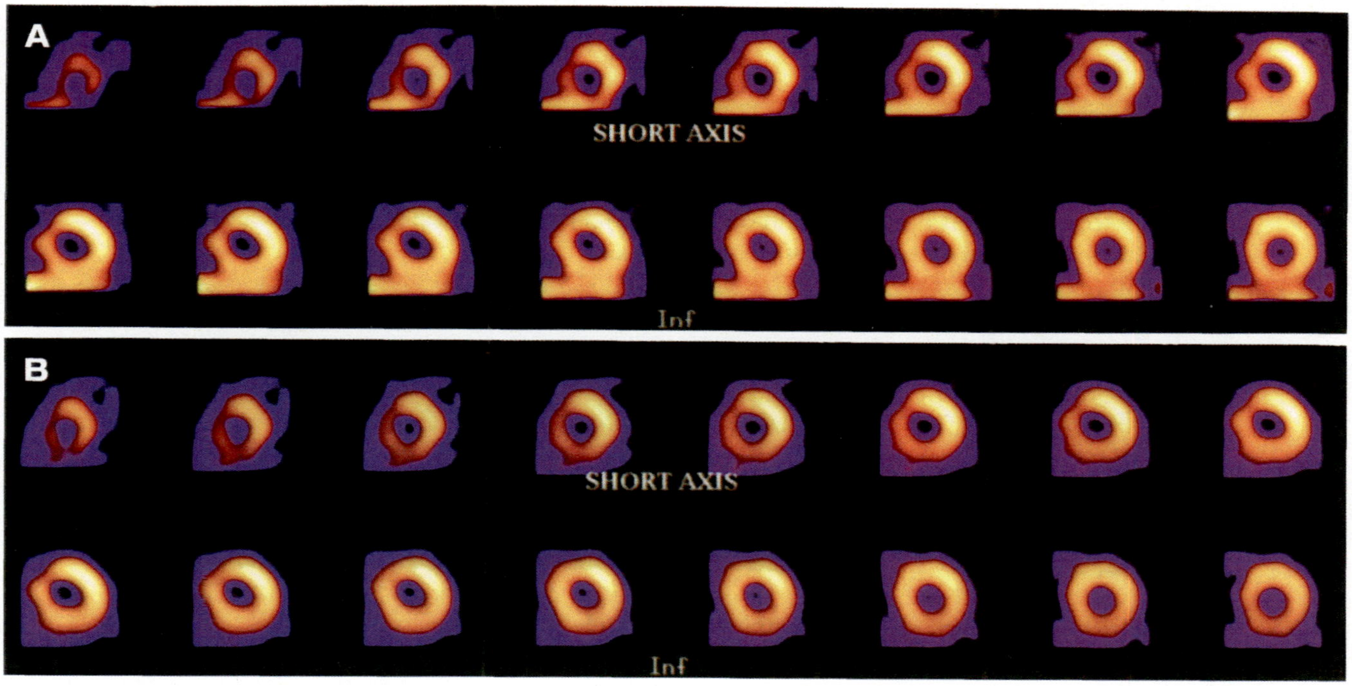

Figure 10.6 SPECT study with bowel activity interference. (A) Short axis slices showing bowel activity interference with the inferior wall. (B) Delayed images acquired after bowel activity has moved.

- Make the patient as comfortable as possible before starting acquisition. Adequate support of the patient's arms and neck will help;
- Explain the procedure and length of study to the patient. Explain the importance of not moving;
- Ask the patient not to talk (carry on conversation) during acquisition;
- The patient should not fall asleep. Snoring or deep breathing could cause a motion artifact; and
- The technologist should stay with the patient and observe him or her for motion during acquisition. If the patient moves, the acquisition can be restarted.

After acquisition is completed and before reconstruction, the acquisition data should be inspected for motion. Motion artifacts may be recognized in several ways:

- Review the SPECT acquisition data in cine for vertical motion. The heart should move back and forth across the screen in a straight line. Look for up or down movement of the heart caused by motion. To help recognize vertical motion and assess its severity, horizontal cursor lines can be bracketed around the heart in the cine display (Fig. 10.9);
- Make a composite image of the SPECT acquisition images. Motion is characterized by a composite image with "tails" or "hurricane sign" (*4*) (Fig. 10.10); and

Figure 10.7 Diagram of misaligned horizontal long axis slice, caused by patient motion.

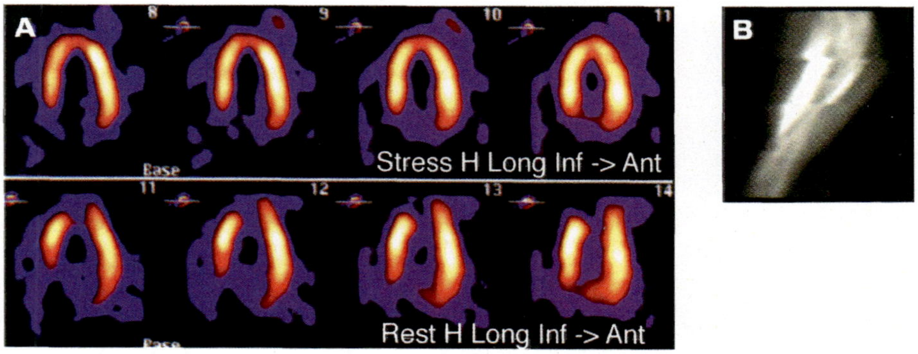

Figure 10.8 Reconstructed patient study with motion artifact. (A) Comparable stress study and rest study horizontal long axis slices. During the rest study the patient moved. (B) The rest study sinogram shows a break resulting from patient motion.

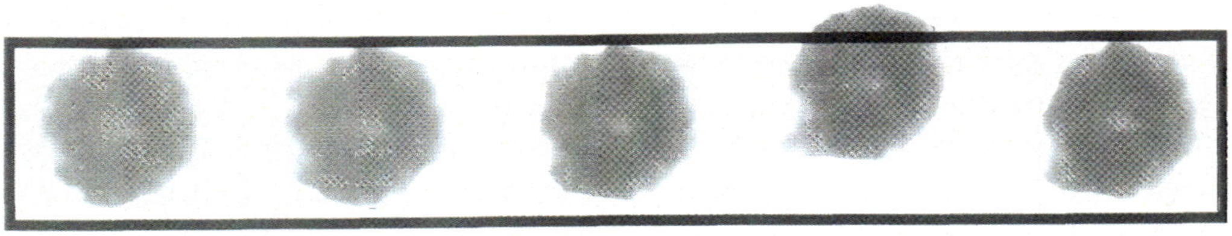

Figure 10.9 Cine display with cursor lines. This diagram shows vertical bouncing of the heart in the cine display as a result of patient motion.

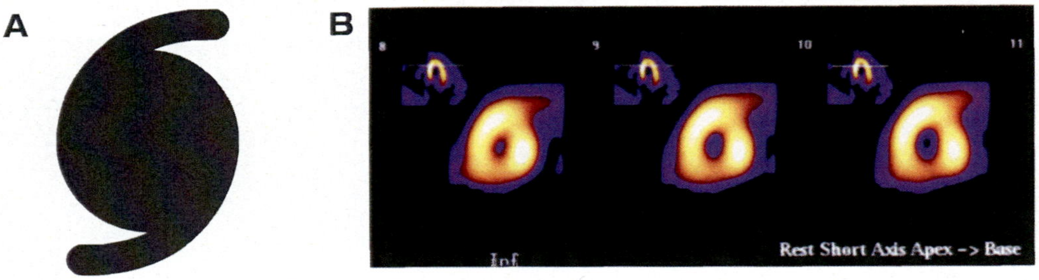

Figure 10.10 Hurricane sign. (A) Diagram. (B) Example scan.

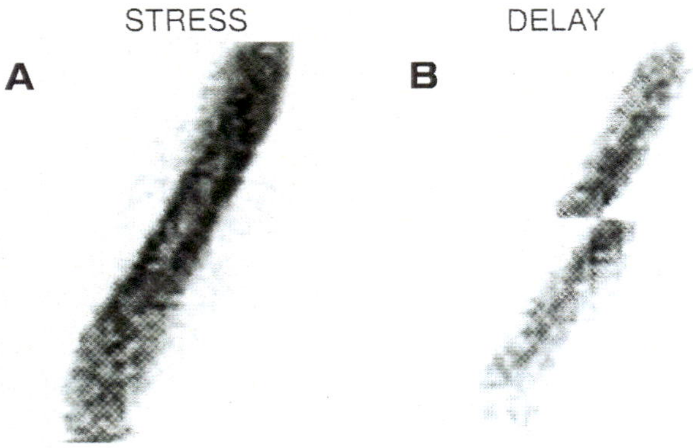

Figure 10.11 Example sinograms. (A) Normal sinogram. (B) Sinogram with a break that resulted from patient movement during acquisition.

- Generate a sinogram from the SPECT acquisition data. A sinogram is a display of a single horizontal profile (line of pixels) from each of the 30 SPECT acquisition projections (Fig. 10.11). A break in the sinogram is evidence of significant patient motion.

Software is available to correct for patient motion (provided the projections in which motion occurs can be identified). Motion correction software has its limitations and cannot correct for all or gross patient movement. It is best to take steps to avoid patient motion. If motion does occur and correction of the data cannot be made, the acquisition could be repeated or the study terminated. A study with motion is technically suboptimal and will be uninterpretable.

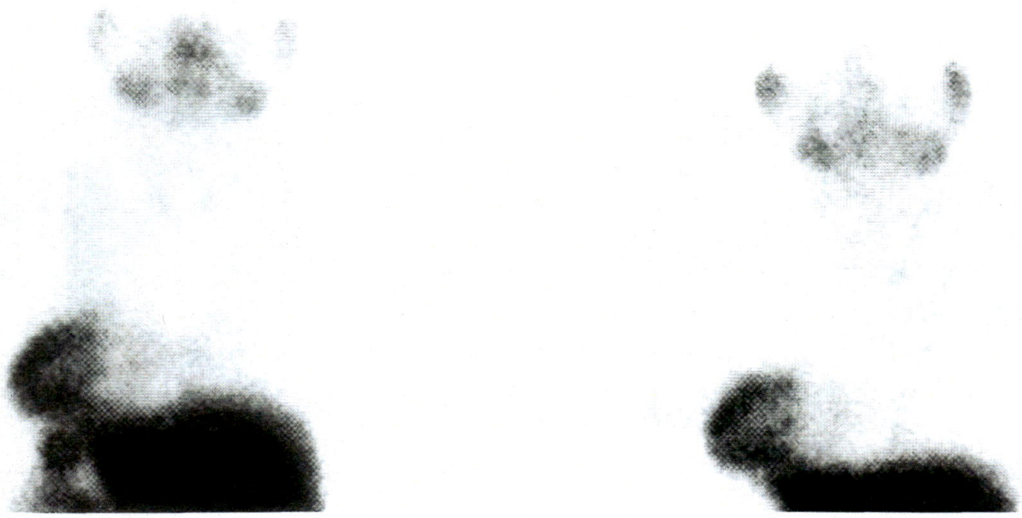

Figure 10.12 Dextrocardia anterior planar study showing the liver on the left side and the heart on the right side.

Other Factors

A number of other factors can affect the reconstructed images and/or the interpretation of the study. Examples include:

- A left bundle branch block (LBBB) can cause a reversible septal defect in a normal left anterior descending coronary artery. A right bundle branch block (RBBB) is less common but may also cause an artifactual reversible septal defect. The patient's stress study electrocardiogram should be inspected, and any BBB should be noted before interpretation of the study (5). It is preferable that a patient with a BBB have a pharmacologic stress study.
- Hot spots also can occur in the images. They may be caused by papillary muscle uptake, localized apical hypertrophy, or by unknown causes. The hot spot intensity may vary in stress and rest images. Normalization of the image to a hot spot will create apparent defects in other myocardial regions. Tomographic slices should be inspected for hot spots and resulting artifacts.
- Apical variations (posterior or anterior) can cause a displacement of the normal apical "dimple," creating a comma-shaped defect. This artifact is most noticeable on polar maps reconstructed from short axis slices. To recognize this artifact, inspect the horizontal long axis slices for a cleft-like defect in the apical region.
- Myocardial hypertrophy can cause reduced coronary blood flow reserve and fixed or partially reversible perfusion defects during exercise. In hypertrophic cardiomyopathy, the septum is usually disproportionately hypertrophied. Image normalization to the relatively hotter septum creates apparent lateral wall defects. The following steps should be taken to recognize this artifact:
  - Obtain a history of hypertension;
  - Review the electrocardiogram and echocardiogram for evidence of left ventricle hypertrophy; and
  - Inspect planar images for increased myocardial mass.
- Rotation of the heart within the thorax alters the normal geometrical relationship of the heart with the detectors in the 180° imaging arc. With levorotation, there is a relative increase in lateral wall count density. With dextrorotation, there is a relative increase in septal wall count density. Patient history and/or physical findings, such as congenital heart disease, chronic obstructive pulmonary disease, or pectus excavatum, may give insight to the recognition of this artifact. Suspicion of artifact from dextrorotation or levorotation should result in inspection of transaxial tomographic slices to determine the rotational axis of the left ventricle; and
- In a small number of patients, the heart will be on the right side of the chest. This normal variant is called dextrocardia. The chambers and great vessels also are inverted right to left. Figure 10.12 shows an example of dextrocardia. Reconstructed images should be labeled in reverse to show the walls accurately (Fig. 10.13).

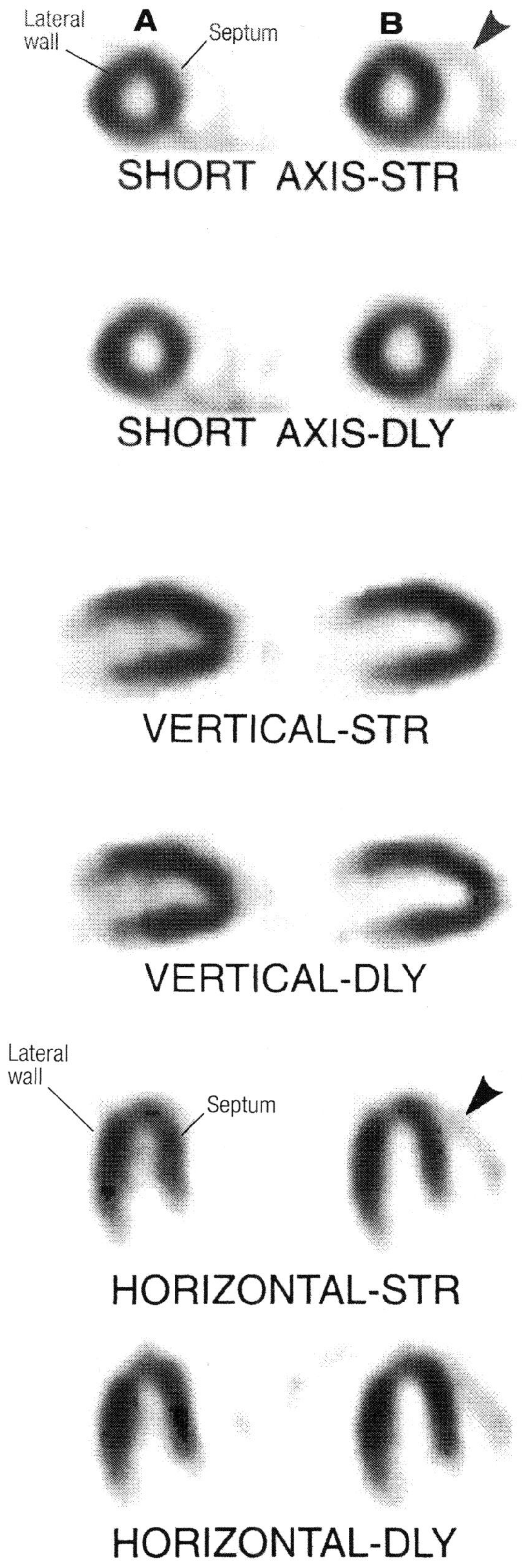

Figure 10.13 Reconstructed SPECT dextrocardia study. Arrows point to the right ventricle, on the opposite side from its normal location. See Figure 9.6 for normal orientation.

References

1. He Zx, Darcourt J, Benoliel J, Migneco O, Magne J, Bussiere-Lapalus F. Major upward creep of the heart during exercise thallium-201 myocardial SPECT in a patient with chronic obstructive pulmonary disease. *J Nucl Med.* 1992;33:1846–1847.
2. Friedman J, Van Train K, Maddahi J, et al. "Upward creep" of the heart; a frequent source of false-positive reversible defects during thallium-201 stress-redistribution SPECT. *J Nucl Med.* 1989;30:1718–1722.
3. DePuey EG, Garcia EV. Optimal specificity of thallium-201 SPECT through recognition of imaging artifacts. *J Nucl Med.* 1989;30:441–449.
4. Sorrell V, Figueroa B, Hansen CL. The "hurricane sign": evidence of patient motion artifact on cardiac single-photon emission computed tomographic imaging. *J Nucl Cardiol.* 1996;3:86–88.
5. DePuey EG, Geurtler-Krawczynska E, Robbins WL. Thallium-201 SPECT in coronary artery disease patients with left bundle branch block. *J Nucl Med.* 1988;29:1479–1485.

Chapter 11

Image Processing/SPECT Study Reconstruction

Cardiac SPECT acquisition data should first be reviewed for patient motion (see chapter 10). If motion is detected, the acquisition data should be motion corrected or the study should be repeated. Preparation for reconstruction, image orientation, filter selection, and attenuation correction are discussed in this chapter.

Image Reconstruction

Before the reconstruction process begins, the technologist must define the reconstruction boundaries in the SPECT acquisition image matrix. A number of projections through the 180° of acquisition should be reviewed to make sure the entire heart is included. The boundaries should include only the heart. Empty space and abdominal structures should be excluded from the reconstruction. Setting boundaries limits the number of images and minimizes the amount of data processed and storage space needed for the study (*1*) (Fig. 11.1)

All acquisition images must be field corrected before backprojection. Field correction for camera nonuniformity is accomplished with a high count density flood source (chapter 8). In addition to field correction, some equipment protocols also smoothe or filter the acquisition data before backprojection.

The process of left ventricle reconstruction begins with the formation of transaxial images that are perpendicular to the long axis of the body. Transaxial images and sagittal images are made in the same way as other SPECT reconstructions. Because the heart sits at an oblique angle in the chest, the transaxial images are not perpendicular to the long axis of the heart. They cut through the heart at an oblique angle. These images are not useful for scan interpretation for several reasons. The myocardial thickness varies with the angle of intersection, which will affect regional count densities. The angle of the heart in the chest varies slightly from patient to patient. Both of these factors minimize the ability to standardize the display and image interpretation.

Image Reorientation

The long axis of the heart must be defined from the standard sectional views so that useful sectional slices of the heart can be created, as discussed in chapter 9. The process of image reorientation involves several steps. First, a transaxial slice through the midportion of the left ventricle that clearly shows the heart cavity is chosen by the technologist processing the study. The display intensity must be adjusted so that the shapes of the heart and cavity are clearly defined. The technologist manually draws a line through the cavity parallel to the long axis of the ventricle. The definition of the long axis is repeated in the same way on a midsection sagittal (or vertical long axis plane) slice (*2,3*). In both cases, the axis drawn must intersect the apex and be parallel to the walls of the heart (Fig. 11.2). The outer confines of the left ventricle walls are defined with cursor lines on both images.

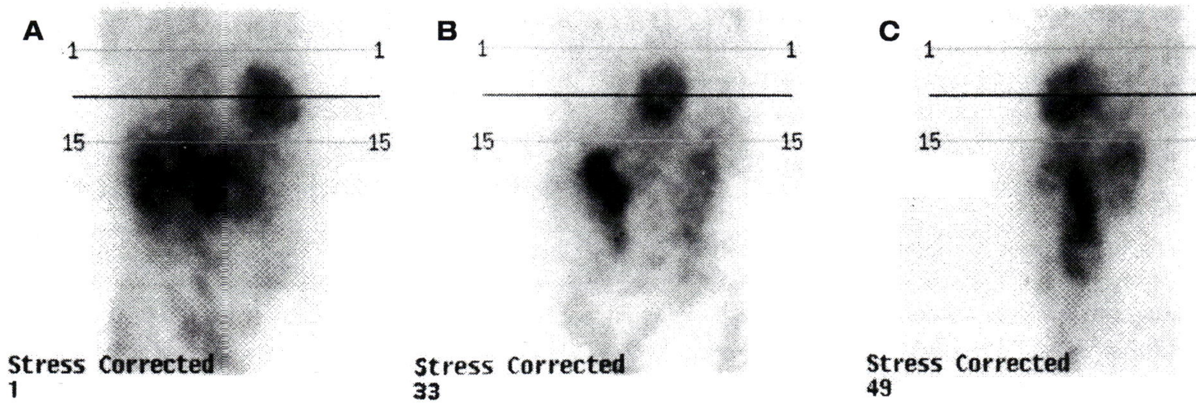

Figure 11.1 Setting boundaries for cardiac SPECT reconstruction.

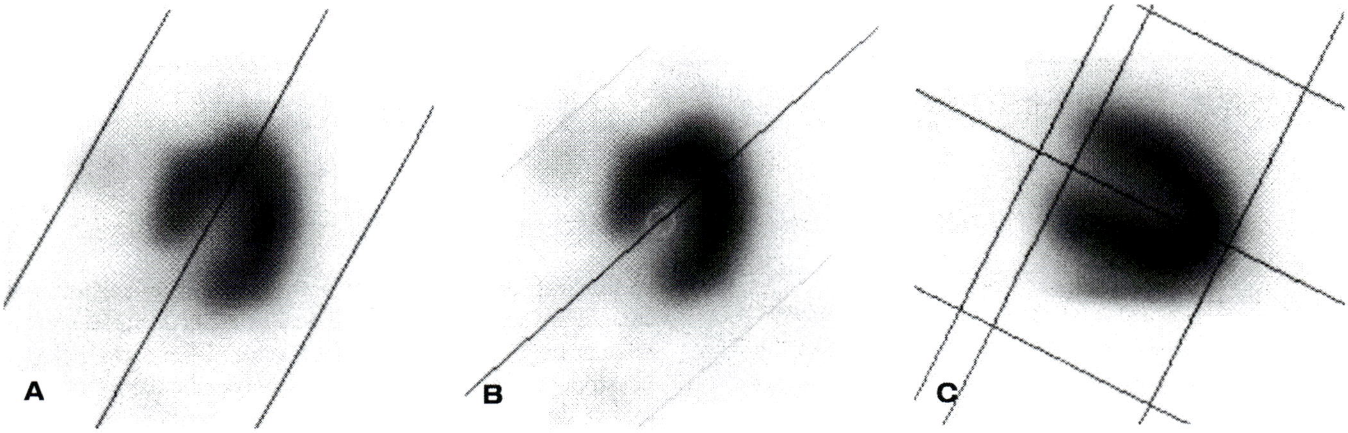

Figure 11.2 Defining the long axis of the left ventricle. (A) Transaxial slice; correct definition of long axis. (B) Transaxial slice; incorrect definition of long axis. (C) Vertical long axis slice; correct definition of long axis.

If the angle of the heart appears to change as the technologist pages through the transaxial slices, a slice should be chosen that is at the midpoint of the angular variations. The angle selection is a visual judgment with manual processing. The heart is usually symmetrical, but in some cases the left ventricle shape is distorted as a result of a peri-infarct aneurysm. In such cases, appropriate angular adjustment must be made to ensure that the reference line defining the long axis correctly intercepts the apex.

Short axis slices of the heart are made by slicing perpendicular to the long axis. The standard views of the heart and the way in which they are displayed are discussed in chapter 9. Stress and rest images will be comparable if the selection of the long axis of the heart is consistent for the two studies. The long axis plane selection of the first study processed should be reviewed before processing the second study.

Manual selection of the long axis of the left ventricle is subject to interobserver variability and is dependent on operator expertise and careful attention in choosing the angle. Correct definition of the long axis of the heart is very important. If it is selected incorrectly, the reconstructed images and/or polar tomograms will be distorted (*4,5*). For this reason, several methods to automate the selection of the long axis and reorientation process have been proposed (*1,2,6,7*). Newer equipment employs automatic processing software. Many of the steps discussed here are automated with the new software.

Filter Selection

Tomographic sectional images are generated by filtered backprojection of image data. Camera manufacturers will make recommendations on filter selection for a given protocol. Pre- and postreconstruction filter selection is based on count density of the study. The count density will vary depending on factors such as the radionuclide used (^{201}Tl or ^{99m}Tc), the dose in millicuries, the matrix size (64 × 64 or 128 × 128), and the protocol (1 or 2 days; single or dual isotope).

Attenuation Correction

Attenuation correction hardware and software are available with newer equipment. Figure 11.3 is a sample study processed with and without attenuation correction. Common methods for application of attenuation correction of SPECT acquisition data do not work well for myocardial perfusion imaging for several reasons. Imaging through 180° rotation does not provide the body contour outline needed. Also, these methods assume a consistent amount of attenuation throughout the volume of the body. Although existing methods of attenuation correction work fairly well in the abdomen, they do not work well in the chest, where tissue composition is less homogeneous (*8*).

Cardiac attenuation correction is accomplished with transmission imaging. A source such as gadolinium-153, americium-241, or barium-133 is used for transmission data (Table 11.1). The emission and transmission studies can be acquired separately or simultaneously. The amount of radiation exposure from the transmission source received by the patient and technologist performing the study is negligible (*9*). In the future, attenuation correction software will be combined with scatter correction and respiratory correction for optimal perfusion images.

TABLE 11.1

Transmission Sources for Cardiac Attenuation Correction

Radionuclide	Half-life	Photon energy (keV)
^{153}Gd	242 days	97,103
^{241}Am	432 years	59 (a decay)
^{133}Ba	10.57 years	Multiple photons, ranging from 30–384 keV

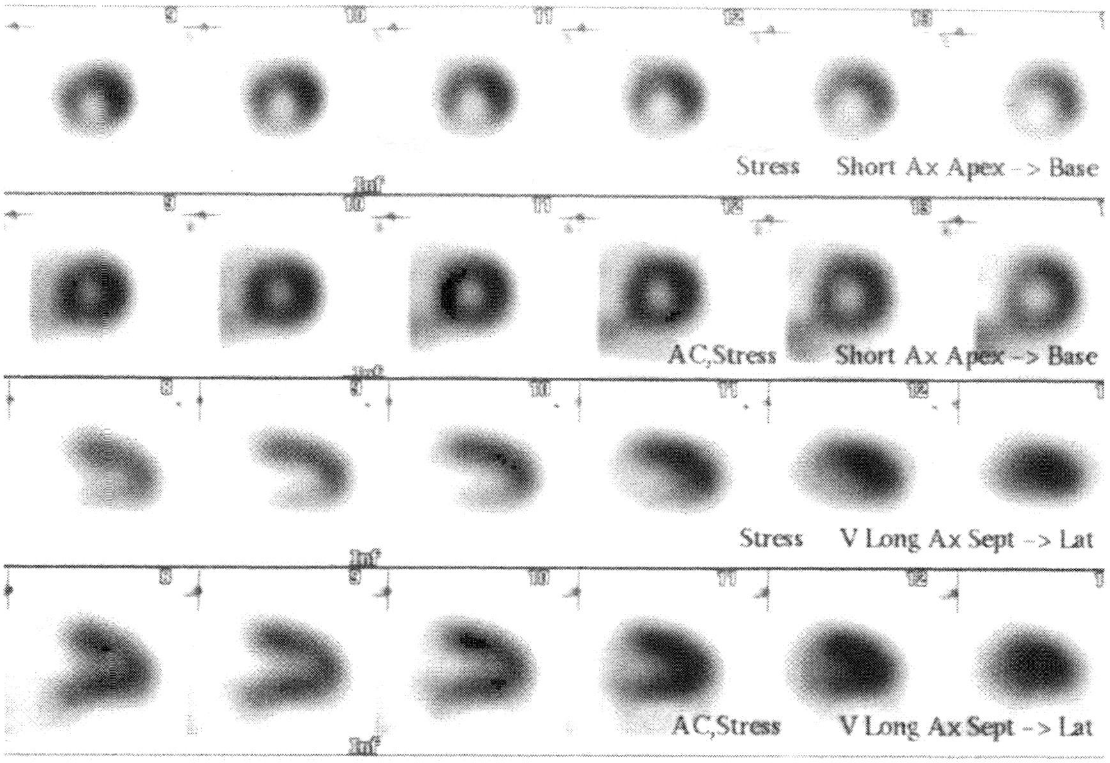

Figure 11.3 Study processed with and without attenuation correction.

References

1. Germano G, Kavanagh PB, Chen J, et al. Operatorless processing of myocardial perfusion SPECT studies. *J Nucl Med.* 1995;36:2127–2132.
2. Germano G, Kavanagh PB, Su HT, et al. Automatic reorientation of three-dimensional, transaxial myocardial perfusion SPECT images. *J Nucl Med.* 1995;36:1107–1114.
3. Germano G, Van Train K, Kiat H, et al. Digital techniques for acquisition, processing and analysis of nuclear cardiology images. In: Sandler MP, Patton JA, Coleman RE, Wacher FJT, Gottschalk A, eds. *Diagnostic Nuclear Medicine.* 3rd edition. Baltimore, MD: Williams & Wilkins; 1996;347–385.
4. DePuey GE, Garcia EV. Optimal specificity of thallium-201 SPECT through recognition of imaging artifacts. *J Nucl Med.* 1989;30:441–449.
5. Lancaster JL, Starling MR, Kopp DT, Lasher JC, Blumhardt R. Effects of error in reangulation on planar and tomographic thallium-201 washout profile curves. *J Nucl Med.* 1985;26:1445–1455.
6. Cauvin JC, Boire JY, Maublant JC, Bonny JM, Zanca M, Veyre A. Automatic detection of the left ventricle myocardium long axis and center in thallium-201 single photon emission computed tomography. *Eur J Nucl Med.* 1992;19:1032–1037.
7. He ZX, Maublant JC, Cauvin JC, Veyre A. Reorientation of the left ventricular long axis on myocardial transaxial tomograms by a linear fitting method. *J Nucl Med.* 1991;32:1794–1800.
8. Cullom SJ, Galt JR, Garcia EV. New approaches to SPECT imaging with scatter and attenuation compensation. In: Iskandrian AE, Verani MS, eds. *New Developments in Cardiac Nuclear Imaging.* New York, NY: Futura Publishing Co., Inc.; 1998:239–264.
9. Christian PE, Tripp DA, Gullberg GT, et al. Patient and occupational radiation exposure from SPECT transmission imaging. *J Nucl Med Technol.* 1995;23:197–201.

Chapter 12

Radiopharmaceuticals and Procedure

Chest pain brings millions of people to hospital emergency rooms each year. Chest pain may be caused by acute myocardial infarction (MI) or unstable angina or may be the result of noncardiac causes. Some but not all patients with acute MI will be easy to diagnose through their symptoms, electrocardiograph (ECG) results, and serum enzyme levels. Nuclear medicine studies are not needed when a diagnosis can be based on the patient's symptoms and conventional tests for MI. Patients with acute MI should be monitored promptly and treated appropriately. For example, maximum myocardial salvage occurs when thrombolytic therapy is started within 6 hours of the acute infarct (*1*).

This chapter will discuss MI-avid imaging with ^{99m}Tc-pyrophosphate (^{99m}Tc-PYP) and ^{111}In-antimyosin. The optimal time for localizing an infarct with these agents is >12 hours after the onset of symptoms. For this reason, the procedures described in this chapter are performed infrequently. In addition, antimyosin antibody may not be available commercially. These agents and procedures would be useful for the evaluation of patients who seek medical attention a few days after the occurrence of symptoms and whose ECG results and/or serum enzyme levels are indeterminate or cannot be interpreted.

Technetium-99m–Pyrophosphate

Description and Mechanism of Localization

^{99m}Tc-PYP was an early bone imaging agent that was found to localize in areas of acute MI. ^{99m}Tc-PYP localizes in the hydroxyapatite crystals in the bone matrix and other areas with crystalline calcium deposits (*2*). Possible mechanisms by which PYP localizes in infarcted myocardium include PYP localization in the calcium deposits in mitochondria and cytoplasm of the infarcted tissue. Also, calcium influx and formation of calcium phosphate complexes are features of damaged muscle cells (*3*).

A ^{99m}Tc-PYP scan may be positive as early as 12 hours after infarct. If a scan is positive as early as 12 hours after an MI, it will become increasingly positive in the subsequent 24–48-hour period (*4*). Uptake of PYP in an acute infarct is also dependent on blood flow to the area and whether or not the region is reperfused or dependent on collateral circulation. In a reperfused MI, the optimal time to image with ^{99m}Tc-PYP is 1–5 days after the coronary episode. For nonreperfused MI, the optimal time to image with ^{99m}Tc-PYP is 2–3 days after the coronary episode (*1*). Once damaged tissue heals, ^{99m}Tc-PYP will no longer localize in the new tissue. The sensitivity of the study decreases sharply after 5–7 days after an acute MI.

Imaging Procedure

The dose of ^{99m}Tc-PYP for infarct imaging is 15–24 mCi, which is administered intravenously. As with other bone imaging agents, patients should be instructed to force fluids (to help clear soft tissue background) and void frequently (to minimize radiation exposure to the bladder). For adequate blood clearance, imaging should be delayed until 3–4 hours after the administration of the dose. All images are acquired with the patient supine. A minimum of three to four static planar images of the chest are acquired with the region of the heart centered in the field of view. These images are an anterior, one or two left anterior obliques, and a left lateral view. An all-purpose or high-resolution collimator is used. Each image should be acquired for a minimum of 500,000–1 million counts. The first image, usually the anterior view, is acquired for counts. Subsequent images should be acquired for the same length of time needed to obtain the desired counts on the first image. Although images may be positive 12 hours to 5 days after the onset of symptoms, the optimal imaging window is 48–72 hours after the episode of symptoms.

SPECT imaging will be particularly useful in differentiating bone pathology from localized uptake in the myocardium and in estimating infarct size. A SPECT study can be acquired in the same way as for myocardial perfusion imaging.

Normal and Abnormal Scan Pattern

The skeleton structures (ribs, sternum, and spine) are well visualized. The kidneys are seen as a result of normal physiological renal excretion. A mild-to-moderate amount of soft tissue background is normal. In normal studies, no focal areas of increased uptake should appear in the region of the heart. Figure 12.1 is an example of a normal ^{99m}Tc-PYP scan.

Increased blood pool activity in the region of the heart appears occasionally as a diffuse circular blush of activity. Increased blood pool activity may be the result of poor renal or cardiac function. Blood pool activity will diminish gradually with time. Depending on the degree of blood

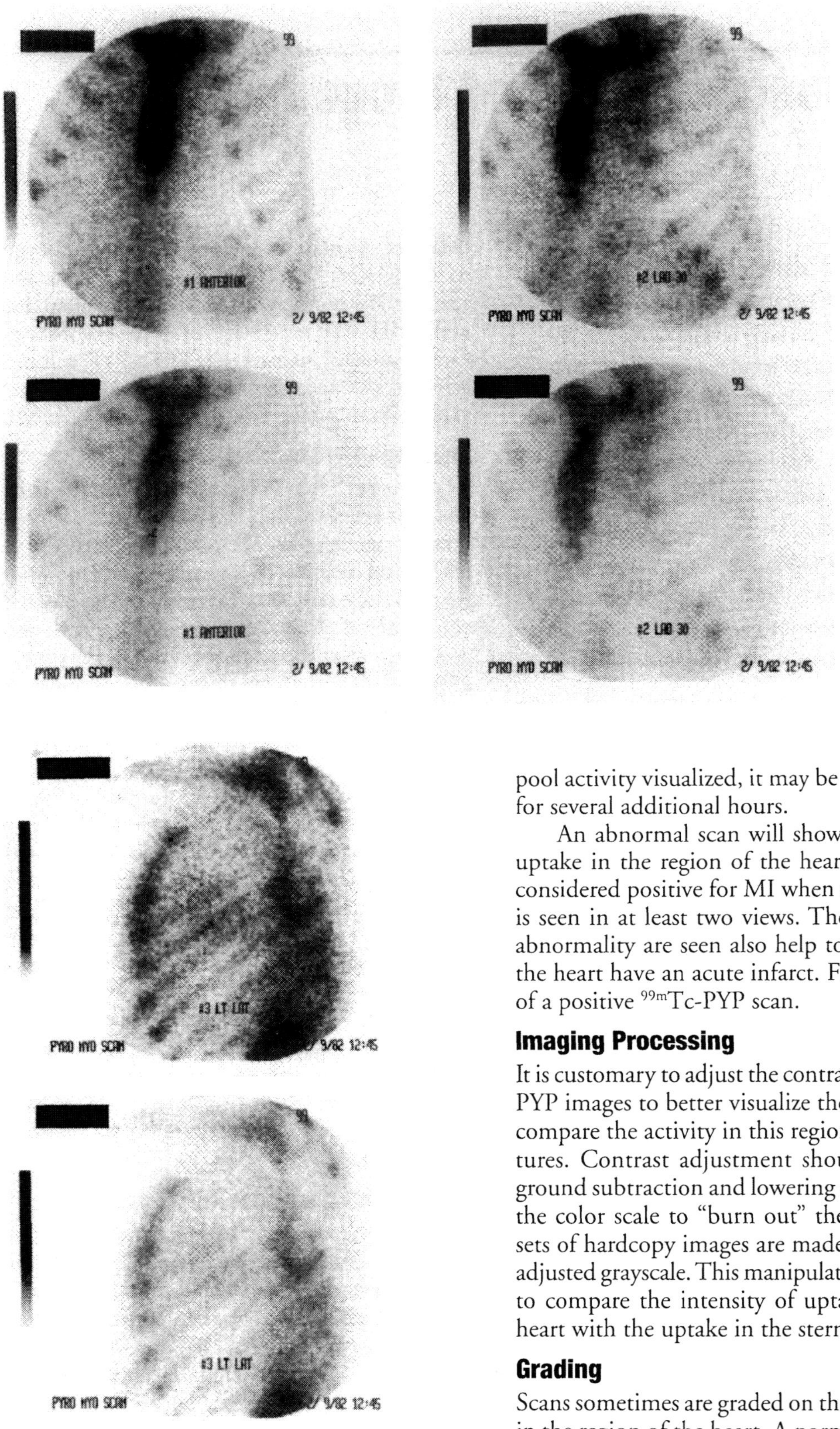

Figure 12.1 Normal ^{99m}Tc-pyrophosphate study. Contrast adjusted. Two intensities displayed.

pool activity visualized, it may be helpful to delay imaging for several additional hours.

An abnormal scan will show focal areas of increased uptake in the region of the heart. A ^{99m}Tc-PYP study is considered positive for MI when a focal uptake of activity is seen in at least two views. The views in which a focal abnormality are seen also help to identify which walls of the heart have an acute infarct. Figure 12.2 is an example of a positive ^{99m}Tc-PYP scan.

Imaging Processing

It is customary to adjust the contrast when reviewing ^{99m}Tc-PYP images to better visualize the region of the heart and compare the activity in this region with the skeletal structures. Contrast adjustment should include some background subtraction and lowering of the upper threshold of the color scale to "burn out" the sternum. Two or three sets of hardcopy images are made with different contrast-adjusted grayscale. This manipulation allows the scan reader to compare the intensity of uptake in the region of the heart with the uptake in the sternum and ribs.

Grading

Scans sometimes are graded on the intensity of uptake seen in the region of the heart. A normal scan would be graded 0. A scan that is considered normal but has faint blood pool activity may be graded 1. Scans with focal increased

TABLE 12.1

Physical Characteristics of ^{111}In

Half-life = 67.92 hours.
Principal photons = 172 keV and 247 keV.

uptake in the region of the heart may be graded 2–4, depending on the intensity of focal activity compared with the ribs and sternum.

Other Causes of Increased Uptake of Technetium-99m–PYP

Bone pathology, inflammation, and calcified soft tissue lesions or tumors in the chest or breast will result in increased uptake of ^{99m}Tc-PYP. Any of these may cause a false-positive scan interpretation. For this reason, patient history is very important.

Myocarditis, pericarditis, acute left ventricular subendocardial MI, stable angina pectoris, ischemic heart disease, postcardioversion status, radiation therapy, and chemotherapy may cause diffuse uptake in the region of the heart.

Focal increased uptake also is seen in acute left ventricular transmural MI, previous MI with persistent activity as a result of myocardial necrosis, unstable angina pectoris with myocardial necrosis, and left ventricular aneurysm.

Indium-111-Antimyosin

Description and Mechanism of Localization

Antimyosin is a Fab fragment monoclonal antibody that binds to the heavy chain of human myosin. Myosin is a contractile protein found in both cardiac and skeletal muscle. Heavy-chain myosin is not found free in the general circulation. If a cell contains myosin and the integrity of the cell membrane is lost (for example, when cell death or injury occurs and the membrane degenerates), the myosin will be exposed to extracellular fluid (*5*). This is the mechanism of localization of antimyosin in acutely infarcted tissue. Although antimyosin can identify acute MI as early as 6 hours after the event, its yield is highest 18–24 hours after infarction. Antimyosin does not localize in normal myocardial tissue or in infarcted tissue more than 2 weeks after the event (*6*). Antimyosin is a specific agent for the detection of necrotic myocardial tissue caused by MI or myocarditis.

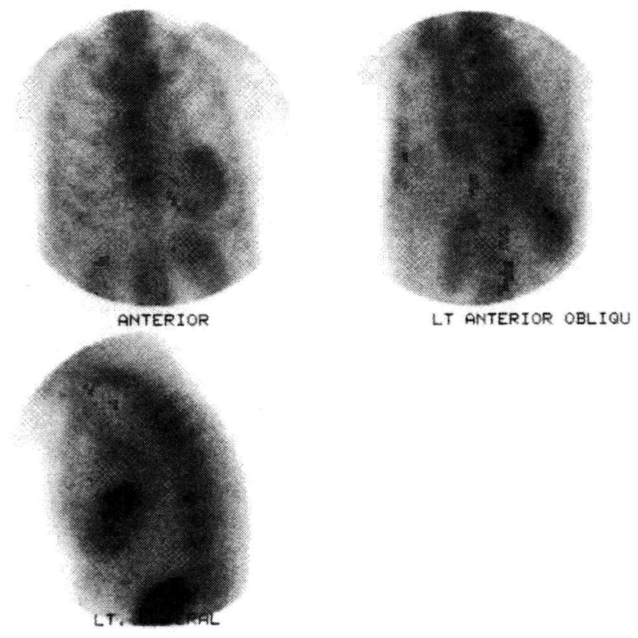

Figure 12.2 Positive ^{99m}Tc-pyrophosphate study. Infarct is seen in the anterolateral region of the heart.

Antimyosin is labeled with ^{111}In. The physical characteristics of ^{111}In are listed in Table 12.1. Monoclonal antibodies are cleared by the liver. The Fab fragment of the antibody is localized in necrotic tissue to a greater degree and is cleared more quickly from the circulation than the whole immunoglobulin. Mean blood clearance at 24 hours is 80%. The antimyosin Fab fragment causes no serious adverse effects.

Imaging Considerations

Images acquired with ^{111}In-antimyosin are similar to those with ^{99m}Tc-PYP. A medium-energy collimator is needed. Images can be acquired as early as 12–24 hours after administration. The target-to-background ratio improves with

TABLE 12.2

Infarct Avid Imaging

Radiopharmaceutical	Mechanism of action	Window of maximum diagnostic yield after infarct	Ideal imaging time after dose administration
^{99m}Tc-pyrophosphate	Complexes with calcium present in infarcted (necrosed) tissue by chemisorption onto calcium hydroxyapatite crystals	48–72 hours	1–3 hours
^{111}In-antimyosin	Binds to the heavy-chain, exposed human myosin	18–24 hours	24–48 hours

time. Images are more often acquired at 24–48 hours after the dose is given to minimize blood pool activity that can obscure a focal defect. Also, the intensity of uptake will vary, depending on the extent of the infarction and whether it is transmural or nontransmural.

Table 12.2 compares the two agents approved for routine use in infarct-avid imaging.

Technetium-99m–Glucarate

A new and promising agent under investigation for infarct-avid imaging is ^{99m}Tc-glucarate (^{99m}Tc-labeled glucaric acid). This agent is useful in the diagnosis of MIs in the acute or hyperacute stage. It is a good agent for MIs of less than 2 days, but its sensitivity decreases after 2–3 days post-MI (*7*).

References

1. Khaw BA. The current role of infarct avid imaging. *Semin Nucl Med.* 1999;29: 259–270.
2. Buja LM, Tofe AJ, Kulkarni PV, et al. Sites and mechanisms of localization of technetium-99m phosphorus radiopharmaceuticals in acute myocardial infarcts and other tissues. *J Clin Invest.* 1977;60:724–740.
3. Parkey RW, Bonte FJ, Stokely EM, et al. Acute myocardial infarct imaging with Tc99m stannous pyrophosphate. In: Serafini AN, Gilson AJ, Smoak WM, eds. *Nuclear Cardiology. Principles and Methods.* New York, NY: Plenum Medical Book Company; 1977:133–143.
4. Willerson JT, Parkey RW, Bonte FJ, Meyer SL, Atkins JM, Stokley EM. Technetium stannous pyrophosphate myocardial scintigrams in patients with chest pain of varying etiology. *Circulation.* 1975;51:1046–1052.
5. Yasuda T, Palacios IF, Dec GW, et al. Indium 111-monoclonal antimyosin antibody imaging in the diagnosis of acute myocarditis. *Circulation.* 1987;76:306–311.
6. Johnson LL, Seldin DW, Becker LC, et al. Antimyosin imaging in acute transmural myocardial infarctions: results of a multicenter clinical trial. *J Am Coll Cardiol.* 1989;13:27–35.
7. Khaw BA, Narula J. New approaches to infarct avid imaging. In: Iskandrian AE, Verani MS, eds. *New Developments in Cardiac Nuclear Imaging.* New York, NY: Futura Publishing Co., Inc.; 1998:171–202.

Chapter 13

Acute Chest Pain Imaging

Millions of people visit hospital emergency departments (ED) each year because of chest pain. In many cases, conventional tests will not confirm or rule out myocardial infarction (MI). Many patients admitted into hospital coronary care units will have noncardiac-related chest pain. A smaller number of patients will be sent home and develop infarction within 48 hours (*1,2*). Chest pain clinics are becoming common adjuncts to ERs for evaluation of patients presenting with angina. It is estimated that more than 10 billion dollars per year are spent to rule out MI in low-risk patients. Myocardial perfusion imaging (MPI) in chest pain clinics has been shown to result in substantial health care dollar savings in this subset of patients (*3*).

Myocardial Perfusion Imaging in Acute Chest Pain

MPI can be used to improve the assessment and management of patients presenting to the hospital ER with chest pain. Adverse cardiac events in patients with normal myocardial perfusion scans is <2%, compared with >70% in high-risk patients with abnormal myocardial perfusion scans (*4*). MPI should be used only when the diagnosis of MI and/or unstable angina is uncertain (for example, when electrocardiography is normal or indeterminate). The patient should have no previous history of MI. If the patient has had a previous MI, the MPI study will be positive and nondiagnostic. The use of MPI in chest pain clinics provides a safe and cost-effective way of managing patients with chest pain.

For acute chest pain imaging, the radiopharmaceutical must be available for injection when the patient with symptoms presents in the ED. Injection of an MPI agent during chest pain provides an opportunity to visualize the blood supply to the different walls of the heart while the patient is symptomatic. If the patient is injected when symptomatic and the images are normal, the chest pain is usually of noncardiac origin and the patient can be sent home. If, on the other hand, the scan is abnormal, the patient should be admitted to the hospital. An abnormal scan may be the result of an MI or unstable angina. The sensitivity of MPI for unstable angina may be decreased if the radiopharmaceutical is not administered during acute pain. Therapeutic interventions during transport to the ER, such as nitroglycerin or other coronary artery vasodilators, also may affect the sensitivity of the test. The patient admitted after positive MPI should have a follow-up scan at a later date when chest pain has subsided. If the abnormality persists, the diagnosis is MI. If an abnormality seen on the earlier study reverses, the chest pain episode was myocardial ischemia. In either case, patient management is optimized. A sample evaluation algorithm for patients presenting to the ED is found in Figure 13.1. An explanation of the algorithm is found in Table 13.1.

Radiopharmaceutical Selection

^{99m}Tc-labeled MPI agents have an advantage over thallium agents for acute chest pain imaging. All MPI agents will distribute in the myocardium proportionally to blood flow. Because of its redistribution property, thallium imaging must begin within 5–10 minutes of injection. Technetium agents do not redistribute and remain fixed in the myocardium. Images with technetium agents can be acquired several hours later, allowing time to stabilize the patient. In addition, technetium agents are readily available 24 hours a day.

Other Considerations

Each facility should develop a policy on the use of MPI imaging in the ED and in triage of patients presenting with acute chest pain. ED personnel should be trained in radiation protection and certified to inject radiopharmaceuticals. The policy should include such things as the selection of patients for MPI, when and how the radiopharmaceutical will be available, who will inject, and when the technologist should be called in to perform the scan.

Advantages

The advantages of myocardial imaging in patients presenting to the ED with acute chest pain include cost savings of health care dollars. Approximately 20% of patients presenting to the ER with acute chest pain will have an acute MI. A significant number of unnecessary admissions for acute MI can be avoided. Moreover, the possibility of missing a diagnosis of acute ischemia is minimized (*5*).

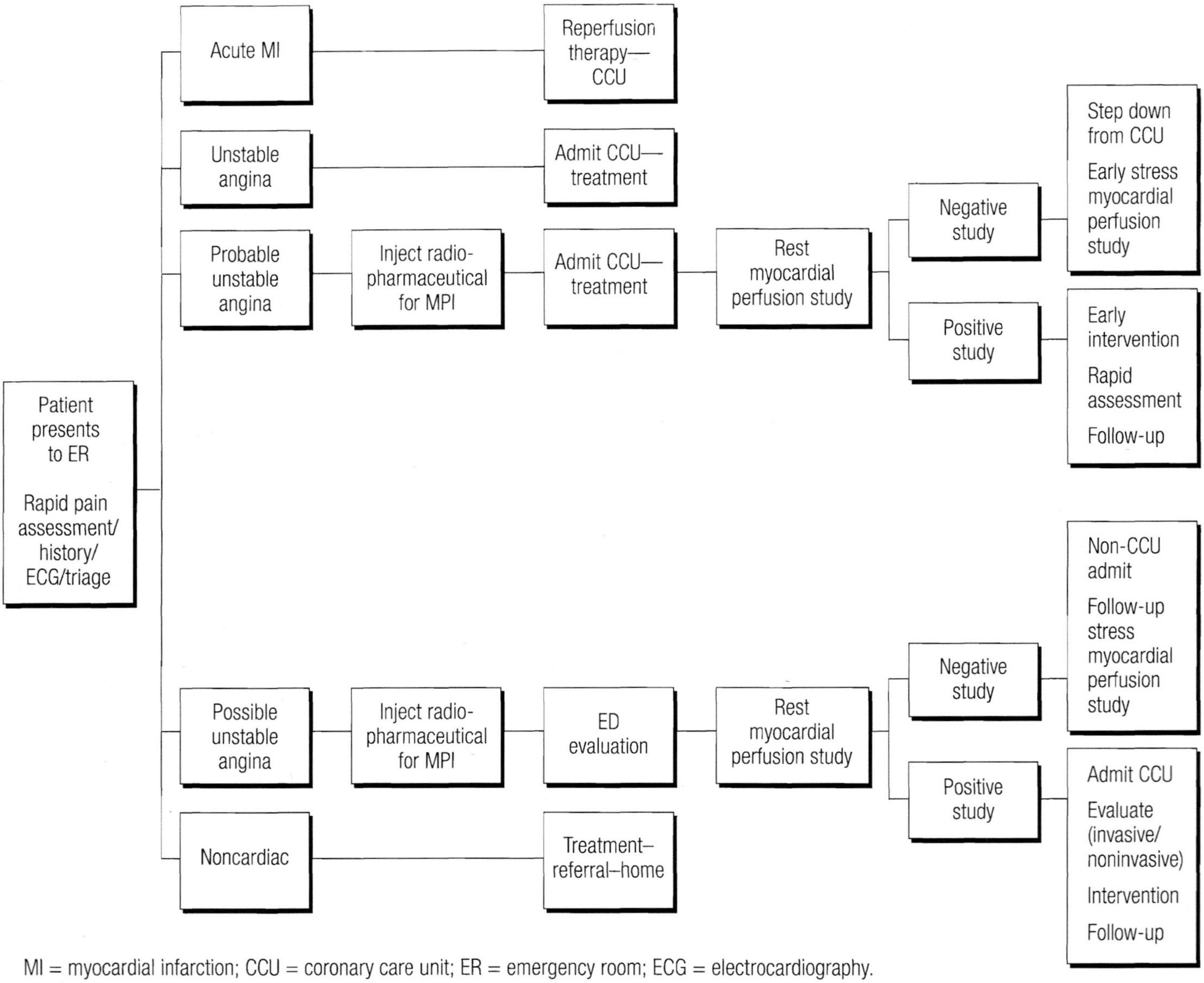

MI = myocardial infarction; CCU = coronary care unit; ER = emergency room; ECG = electrocardiography.

Figure 13.1 Sample evaluation algorithm for patients presenting to the emergency room with chest pain.
Adapted, with permission, from DuPont Pharmaceuticals Company.

TABLE 13.1

Understanding the Sample Evaluation Algorithm (Figure 13.1)*

ED presentation	Probability of AMI	Diagnostic criteria	Diagnostic strategy
Acute MI	High (>95%)	■ Ischemic ST ↑ ■ Signs of acute posterior MI ■ Hyperacute T-wave changes ■ LBBB new or of unknown age	Serial ECGs, enzymes
Unstable angina	High (50%–95%)	■ Ischemic ST ↓ ■ ST-T abnormalities	Serial ECGs, enzymes
Probable unstable	Moderate (5%–50%)	■ Typical prolonged (>30 minutes) symptoms, normal or nondiagnostic ECG ■ Atypical prolonged symptoms and nondiagnostic ECG ■ New onset CHF unexplained by other etiology	Serial ECGs, angina enzymes, ST-segment monitoring, imaging with myocardial perfusion agent
Possible unstable angina	Low (<5%)	■ Typical but not prolonged (<30 minutes) symptoms, normal or nondiagnostic ECG ■ Atypical but not prolonged symptoms, nondiagnostic ECG	Imaging with myocardial perfusion agent
Noncardiac	Very low (<1%)	■ Uncharacteristic history ■ Normal or nonischemic ECG	Discretionary

* Adapted, with permission, from DuPont Pharmaceuticals Company.

ED = emergency department; AMI = acute myocardial infarction; MI = myocardial infarction; ECG = electrocardiogram; LBBB = left bundle branch block; CHF = congestive heart failure.

Disadvantages

The disadvantages include the logistics of providing this service. It may be difficult to provide the radiopharmaceutical and personnel coverage during nonroutine working hours. Also, the diagnosis of noncardiac causes of chest pain may be overlooked inadvertently, such as in cases of dissecting aortic aneurysm, pericarditis, or pulmonary embolism.

References

1. Varetto T, Cantalupi D, Altieri A, Orlandi C. Emergency room technetium-99m sestamibi imaging to rule out acute myocardial ischemic events in patients with nondiagnostic electrocardiograms. *J Am Coll Cardiol.* 1993;22:1804–1808.
2. Hilton TC, Thompson RC, Williams HJ, Saylors R, Fulmer H, Stowers SA. Technetium-99m sestamibi myocardial perfusion imaging in the emergency room evaluation of chest pain. *J Am Coll Cardiol.* 1994;23:1016–1022.
3. Tatum JL, Jesse RL, Kontos MC, et al. Comprehensive strategy for the evaluation and triage of the chest pain patient. *Ann Emerg Med.* 1997;29:116–125.
4. Hilton TC, Fulmer H, Abuan T, Thompson RC, Stowers SA. Ninety-day follow-up of patients in the emergency department with chest pain who undergo initial single-photon emission computed tomographic perfusion scintigraphy with technetium 99m-labeled sestamibi. *J Nucl Cardiol.* 1996;3:308–311.
5. Heller GV, Stowers SA, Hendel RC, et al. Clinical value of acute rest technetium-99m tetrofosmin tomographic myocardial perfusion imaging in patients with acute chest pain and nondiagnostic electrocardiograms. *J Am Coll Cardiol.* 1998;31:1011–1017.

Chapter 14

Indications for Dynamic Cardiac Imaging

The heart is a hollow muscular organ. (Cardiac anatomy was discussed in chapter 1.) The rhythmic contractions of the heart pump blood through the great vessels to all parts of the body. At rest and for regular everyday activities, the heart contracts between 60 and 100 times per minute. The average heart rate (HR) is 72 beats per minute. The normal heart contracts with a regular rhythm, meaning that the length of each contraction is the same. A variety of factors affect both HR and rhythm. As the demand on the heart is increased, the HR will go up. Physical and emotional stress causes an increase in HR. Illness or medications can cause an increase or decrease in HR.

Heart Muscle Innervation

Conduction System

The intrinsic mechanism by which the heart contracts is called the conduction system of the heart. Contraction is controlled by specialized cells in the heart that generate and distribute electrical impulses. The impulses stimulate cardiac muscle fibers to contract. A group of cells located in the right atrial wall inferior to the opening of the superior vena cava initiates each heart contraction. This group of cells is called the sinoatrial (SA) node. The SA node is also referred to as the "natural pacemaker" of the heart. The cells within the SA node spontaneously depolarize, triggering contraction. The impulses of the SA node spread across the atria and stimulate contraction. Contraction begins in the upper part of the atria and spreads toward the atrioventricular (AV) valves. This action helps move blood from the atria into the ventricles.

Within the interatrial septum, near the junction of the atria and ventricles, is the AV node, another specialized group of cells. Adjacent to the AV node is the AV bundle, a group of specialized cells that passes into the interventricular septum. The AV bundle is also known as the "bundle of His." In the superior portion of the interventricular septum, the bundle of His divides into right and left bundle branches. Each branch travels down the septum. From the apex of the heart, branches called the Purkinje fibers traverse to all ventricular cells as they travel back toward the base of the heart (Fig. 14.1).

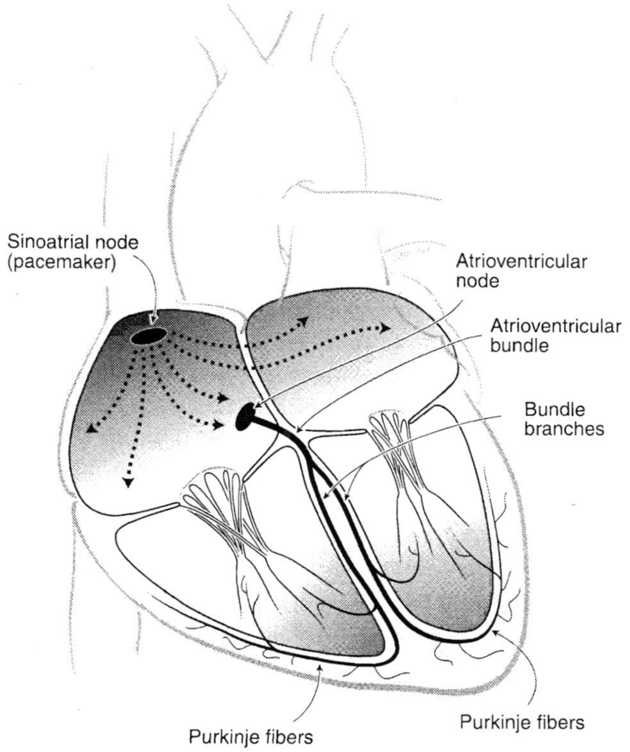

Figure 14.1 Conduction system of the heart.

The AV node is one of the last areas in the atria to receive impulses from the SA node. When the impulses reach the AV node, depolarization of the node occurs and contraction spreads via the bundle branches and Purkinje fibers throughout the ventricle. There is a short delay of approximately 0.10 seconds between atrial contraction and the initiation of ventricular contraction.

Electrical impulses that travel through the conduction system of the heart and electrical changes that occur during the cardiac cycle can be monitored and recorded by electrocardiography (ECG). An understanding of the basic ECG tracing is pertinent to the dynamic cardiac procedures covered in this section. The basics of ECGs are covered in the final section of this book.

Parameters of Function

Dynamic cardiac imaging techniques provide a means of assessing the functional capacity of the heart. Cardiac wall motion is assessed visually. The walls of the ventricles normally contract at the same time and rate in a synchronous

TABLE 14.1
Terminology

ED = end diastolic volume
ES = end systolic volume
SV = stroke volume = ED – ES
%EF = % ejection fraction = SV/ED × 100
LVEF = left ventricular ejection fraction
RVEF = right ventricular ejection fraction

fashion. The extent of contractions should also be noted. Visual evaluation is somewhat subjective and requires an experienced eye. Adjusting the speed of the cine display can be helpful. Computer processing techniques can provide additional information (see chapter 17).

A basic measure of cardiac function is the ejection fraction (EF). Right and/or left ventricle EF can be determined. Each ventricle's EF can be calculated by dividing its stroke volume (SV) by its end diastolic (ED) volume. The SV is the difference between the ED and end systolic (ES) volumes (see EF-related terms in Table 14.1). Normal EF values are listed in Table 14.2. These are the values expected at rest and during normal, everyday activities. Physical exercise should cause an increase in EF.

Cardiac Pathophysiology

Myocardial Infarctions and Resulting Aneurysms

Many conditions can affect the functional capacity of the heart. Arteriosclerotic heart disease and myocardial infarction (MI) were described in chapter 1. Scar tissue resulting from an MI cannot contract. A region of scar tissue in the heart can interrupt the normal conduction system. Depending on the location of the infarct, the resulting scar tissue may become stretched or dilated and, over time, form a sac or pouch as a result of the force of contraction of the rest of the heart muscle. This type of dilatation is known as a ventricular aneurysm. An aneurysm is more likely to develop if the infarct was at the apex or inferior wall. An aneurysm will exhibit paradoxical movement, meaning it will bulge out when the ventricle is contracting. Dyskinesia is another term that describes the movement of a ventricular wall with an aneurysm. An aneurysm can significantly reduce a ventricle's EF. An aneurysm can be corrected surgically. If an aneurysm goes undetected, it may rupture.

Valvular Malfunctions

Valves in the heart that do not function properly can cause regurgitation of blood, that is, movement of blood from a ventricle back into the atria. Valvular stenosis can restrict the flow of blood and limit the amount of blood delivered to a ventricle and/or affect the EF of that ventricle. Heart

TABLE 14.2
Normal Ejection Fraction Values

Left ventricular ejection fraction = 50%–80%
Right ventricular ejection fraction = 40%–60%

valve problems can result from congenital malformations, endocarditis (an infection of the lining of the heart), or rheumatic fever.

Myocarditis

Myocarditis is an infection of the myocardium and can be caused by a bacteria, virus, or parasite. Although curable, myocarditis may damage the overall functional ability of the heart.

Heart Block and Bundle Branch Block

Various types of aberrant conduction, such as heart blocks and bundle branch blocks, can affect cardiac function. These are defined and described in the final section of this text.

Effects of Aging on the Heart

Aging is accompanied by a gradual decline in the strength of the contraction of the heart and the volume of blood it is able to pump each time it contracts. This decline is partly a result of the deconditioning that comes from a lack of physical activity with increasing age. The decline in heart function is also the result of a reduction in the ability of the body to replace old cells as they die (including cardiac muscle cells) and a reduction in the capillary vasculature. A regular exercise program can improve heart function by promoting the development of the collateral vessels and maintaining adequate blood supply.

With aging, both the endocardium and conducting system of the heart are affected by increased deposition of connective tissue. The deposition of calcium can interfere with heart beat initiation and/or regulation, leading to arrhythmias and/or heart blocks. The heart valves also tend to become more rigid as a result of calcium deposits, which may cause functional impairments.

The size of the heart does not change significantly with age unless disease is present. An enlarged heart is often the result of ischemic or dilated cardiomyopathies, congestive heart failure, or other cardiac-related problems.

Indications for Dynamic Cardiac Imaging

Dynamic cardiac imaging is useful in assessing the heart's ability and capacity to function as a pump. It is also useful in the detection and evaluation of cardiac functional impairment and in the management of patients with known cardiac disease (*1*). Indications for dynamic cardiac imaging are listed in Table 14.3. Wall motion abnormalities are diagrammed in Figure 14.2.

TABLE 14.3

Indications for Dynamic Cardiac Imaging

1. Assess either left or right ventricular ejection fraction (EF) at rest or during exercise.
2. Evaluate regional wall motion abnormalities, such as hypokinesia, akinesia, and dyskinesia (aneurysm), during rest or during exercise.
3. Evaluate ventricular function (EF and wall motion):
   - Before cardiac catherization;
   - After a myocardial infarction;
   - In patients with valvular disease;
   - To monitor patients on doxorubicin and related chemotherapy agents; and
   - To evaluate responses to therapies, including exercise, angioplasty, open heart or cardiac bypass surgery, and others.
4. Estimate cardiac output, ventricular volumes, and other cardiac functional parameters.
5. Assess diastolic function of the left ventricle, such as peak filling rate and time to peak filling rate.
6. Evaluate and quantify left-to-right or right-to-left cardiac shunts.

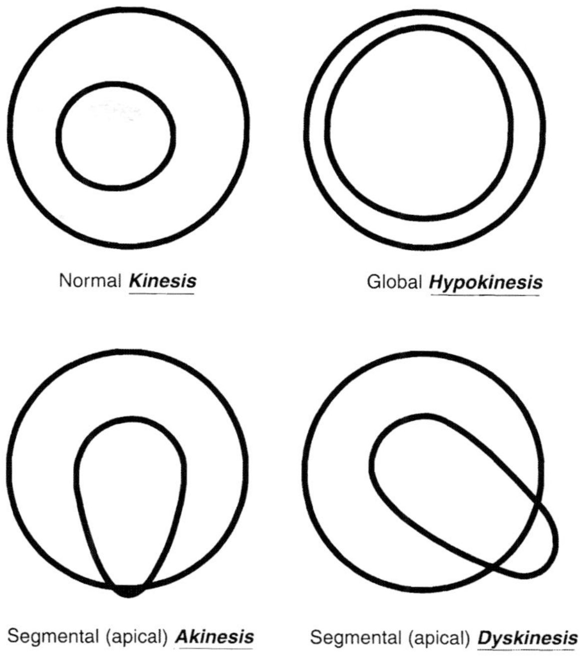

14.2 Wall motion terms.

Dynamic cardiac imaging techniques also can be used in the detection and evaluation of cardiac shunts. This application will be discussed briefly in chapter 20.

Cardiotoxic Drugs

Some chemotherapy agents, particularly the anthracycline antineoplastic drugs, are known for cardiotoxicity. Doxorubicin is an example of a commonly used chemotherapy anthracycline derivative. Because of the untoward side effects of this class of drugs, there are limitations to its use. Usually, no significant toxicity occurs below cumulative doses of 350 mg/m^2 of these chemotherapy agents. With doses greater than 550 mg/m^2, approximately one-third of patients develop cardiotoxicity.

Serial measurements of EF are needed to monitor chemotherapy-induced cardiotoxicity and appropriately manage the patient. A baseline study of left ventricular function should be obtained before a patient starts receiving any cardiotoxic chemotherapy agent. Patients with a baseline EF of less than 30% should not commence treatment with anthracycline derivatives. Patients who are treated with these agents should have their EF checked when the cumulative dose of chemotherapy is 300 mg/m^2. Beginning at 350–450 mg/m^2, the EF should be checked before each chemotherapy dose. Chemotherapy should be discontinued when the EF drops below 40%. If a single dose of chemotherapy at any time causes a decrease in EF of more than 10%, the therapy should be discontinued (*2,3*).

References

1. Port SC, Wackers FJ. Clinical application of radionuclide angiography. *J Nucl Cardiol.* 1995;2:551–558.
2. Shuman RD, Ettinger DS, Abeloff MD, Livengood SV, Fortuin NJ. Comparative analysis of noninvasive cardiac parameters in the detection and evaluation of adriamycin cardiotoxicity. *Johns Hopkins Med J.* 1981;149:57–63.
3. Alexander J, Dainiak N, Berger HJ, et al. Serial assessment of doxorubicin cardiotoxicity with quantitative radionuclide angiocardiography. *N Engl J Med.* 1979;300:278–283.

Chapter 15

Dynamic Cardiac Imaging Method Options

Dynamic cardiac imaging can be accomplished by the gated equilibrium blood pool method and/or the first-pass technique. Both can be performed with the patient at rest or during exercise.

Gated Equilibrium Blood Pool Method

The gated equilibrium blood pool method is also referred to as multiple gated acquisition or MUGA. In this method, data from many consecutive cardiac cycles are overlapped. The radiopharmaceutical equilibrates in the patient's blood pool before acquisition. Multiple views of the heart can be obtained with a single dose.

In this method, the cardiac cycle is divided into a predetermined number of frames or images. The number of frames or images of the cardiac cycle acquired generally ranges from 16 to 32 and sometimes more. An acquisition matrix for each frame is made in the computer memory and retained there until the acquisition of the desired counts density is complete.

The time per frame (or image) is determined by dividing the patient's average cardiac cycle by the number of frames desired. The average adult heart rate (HR) is 60–80 beats per minute. The average adult cardiac cycle, therefore, is approximately 0.75–1.0 second in length. For example, assuming a normal sinus rhythm of 60 beats per minute and a 24-frame acquisition, the time per frame would be 42 milliseconds (1 second/24 frames).

The patient is connected by three leads to an electrocardiograph (ECG), which is also interfaced to the imaging system. During acquisition, each time the computer sees an R wave, it will start to acquire data beginning at frame 1 for the predetermined millisecond interval. It will continue to acquire data in each consecutive frame until the next R wave occurs, at which time the acquisition will be redirected back to frame 1. This is referred to as "gating." It is as if a pointer or triggering mechanism in the computer memory directs acquisition of data into the individual frames. Each time an R wave is detected, the pointer moves to image 1. From image 1, the pointer moves to subsequent frames at every predetermined millisecond time interval. By the time the pointer reaches the last frame and the millisecond time interval is completed, the computer should receive another R wave redirecting the pointer and acquisition back to image 1. The process will continue until the desired information count density has been acquired. Data from many cardiac cycles are overlapped this way to make a composite picture of the heart in different stages of contraction and relaxation (Fig. 15.1).

Multiple Gated Acquisition Validity

The data acquired in each frame must come from exactly the same part of each cardiac cycle. Otherwise, a true representation of heart contraction and relaxation will not be obtained. The R wave represents end diastole. Usually the first image will be a representation of end diastole. All subsequent images will represent what follows end diastole during the average cardiac cycle. Care must be taken to ensure that other waves or signals are not misinterpreted as an R wave by the computer. This would result in acquisition of data into the wrong frames. Gating sometimes may occur incorrectly on the P or T wave, pacemaker spike, muscle artifact, or other artifactual signals. Avoidance of such artifacts is discussed in chapter 17.

Erroneous acquisition also can result from fluctuations in beat length or R-to-R interval. Variations in beat length affect the diastolic more than the systolic portion of the cycle. If an R wave is detected, the pointer and acquisition will be redirected back to image 1, regardless of the pointer's location. If the beat length (R-R interval) is shorter than average, data will not be acquired into one or more of the latter frames. If the R-R interval is longer than average, data acquisition will pause and the computer will wait for the next R wave before moving the pointer back to image 1 (Fig. 15.2). The frames at the end of the cycle represent the heart when it is near end diastole. Acquisition of data from beats that are shorter or longer than average will result in a falsely diminished number of counts in latter frames. This is referred to as frame drop-off. As a result, the more arrhythmic the heart rate, the less accurate is the representation of diastolic function.

Because ejection fraction (EF) is calculated from the systolic portion of the cycle, it is less affected by beat length than are other parameters relating to the diastolic portion of the cycle, such as peak filling rate and time to peak filling rate. To be valid for systolic and diastolic analysis, MUGA requires a constant regular HR and R-R interval. Even with normal sinus rhythm, there will be subtle varia-

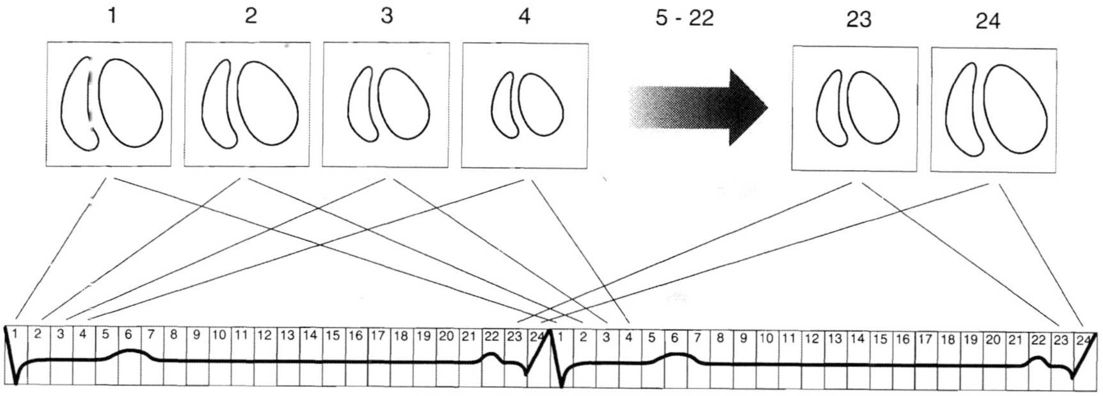

Figure 15.1 Diagram of multiple gated acquisition, normal sinus rhythm. The average cardiac cycle is divided evenly into a number of segments. The time per segment is fixed. An acquisition matrix or frame for each segment is constructed in the computer memory. Data from many cardiac cycles are overlapped into each frame.

tions in the R-R interval. The beat length can be known only after it is completed and the next R wave is detected. "Bad" beats must be excluded from the acquisition. This is accomplished in several ways.

Handling Arrhythmias

The average R-R interval is computed first. A tolerance limit in milliseconds or percentage of the R-R interval (generally ±10%) is chosen. The wider the tolerance limit, the more irregular beats will be acquired. Beats outside of the tolerance limit can be discriminated against in two ways.

Post-Beat Filtration

In the first method, the length of an R-R interval is evaluated after the counts from that particular cycle are applied to the acquisition in memory. If the R-R interval falls within the tolerance limit, data collection of the next cardiac cycle continues. If the length of the R-R interval is not within the tolerance limit, acquisition pauses until an acceptable R-R interval is detected. At that point, acquisition resumes. The postbeat filtration method was developed because beats after a premature ventricular contraction have been shown to have an abnormal degree of contractility, which will subsequently cause a distortion of the volume curve generated from the study. One major disadvantage of this method is that data from some bad beats will be collected. Also, the last frame of the study will contain fewer counts because of short beats.

Buffered Bad Beat Rejection

In the second method, data from each beat are held in a buffer in the computer memory until the end of the beat. The beat length is then evaluated to determine if it falls within the tolerance limit. The data will be applied to the acquisition or discarded, depending on whether or not it is a valid beat. Buffered bad beat rejection eliminates all bad beats in terms of beat length. This method requires more computer memory.

With a stringent tolerance limit (e.g., ±10%), the more arrhythmias, the longer the acquisition time. Also, the more beats rejected or eliminated from the acquisition, the less representative are the data of the average cardiac cycle.

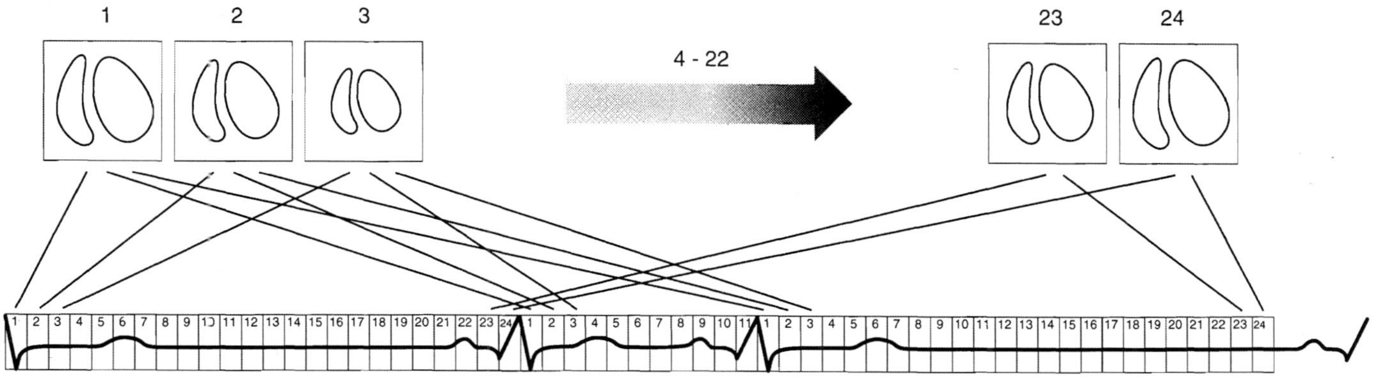

Figure 15.2 Diagram of multiple gated acquisition, irregular heart rate. The R wave directs acquisition of data into image 1. All frames will not receive data from short beats. Data at the end of longer than average beats are lost.

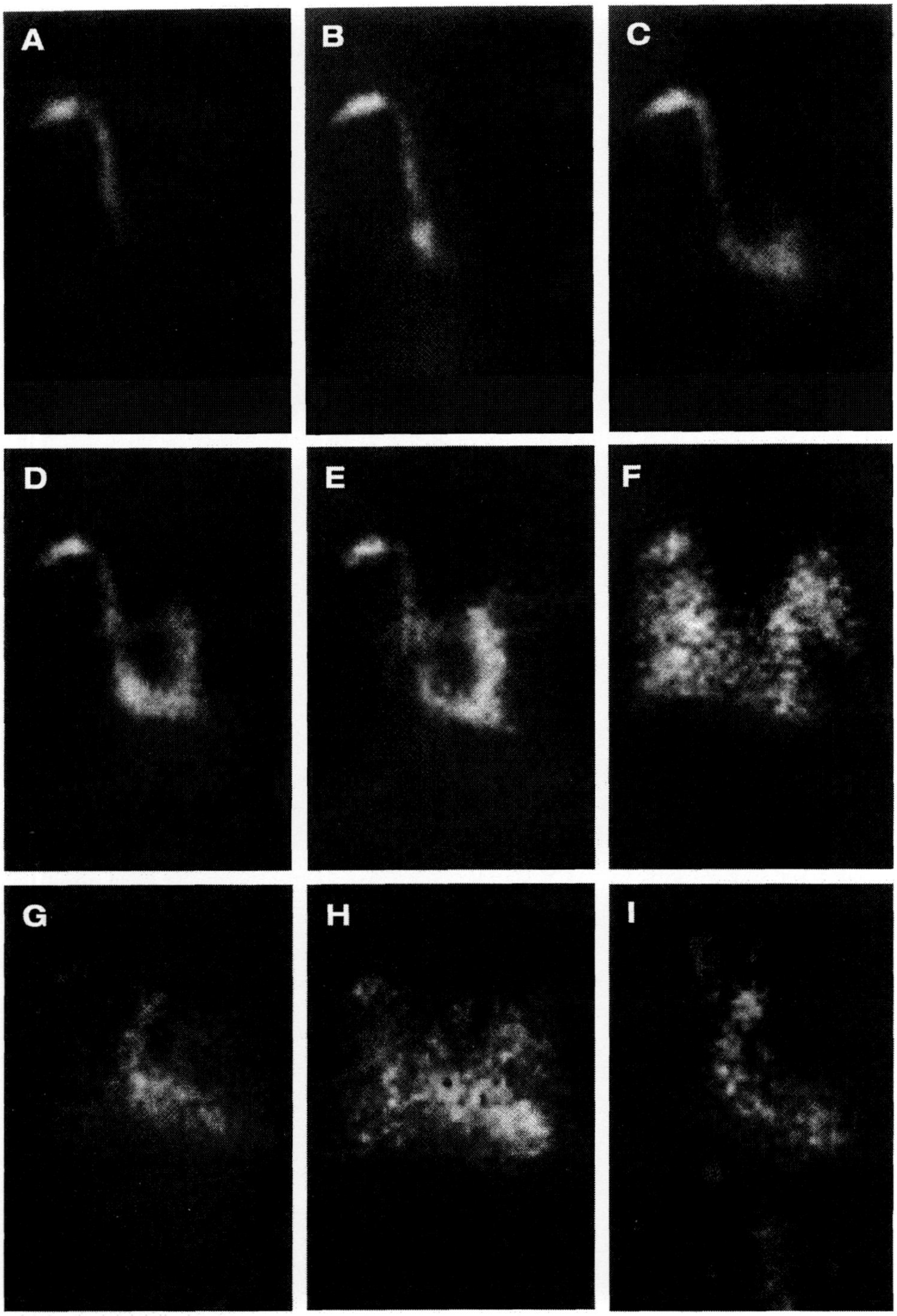

Figure 15.3 Example of first-pass study. Selected frames from a first-pass study showing the bolus traveling through the: (A) superior vena cava; (B) right atrium; (C, D) right ventricle; (E) pulmonary outflow tract; (F) pulmonary circulation; (G, H) left heart; and (I) left heart and aorta.

List Mode Multiple Gated Acquisition

For list mode MUGA, the radiopharmaceutical must have equilibrated in the patient's blood stream, as in frame mode MUGA. In list mode (also called serial mode), counts from the camera are not processed into any image matrix during acquisition. Instead all counts are recorded in sequential order on a disk. Time markers from 1–10 milliseconds apart are recorded with the count data. The R waves are also recorded in the sequential list of data. The acquisition requires 6–8 million storage locations or approximately 12 megabytes of disk space.

Once the acquisition is complete, the data can be selectively reformatted into images. The length of each R-R interval in the list can be evaluated. Data from beats with similar length can be formatted into a frame mode MUGA study. One or several frame mode MUGA studies can be generated from one list mode MUGA. For example, a separate frame mode MUGA study for two distinct beat lengths can be generated from list mode data for a patient with bigeminy.

The advantage of list mode is that the data can be analyzed and reformatted selectively in different ways. Forward/backward framing or gating also can be used in the reformatting process. Forward gating is the standard gated acquisition method. Forward/backward gating involves constructing two image sequences: one forward from the R wave and one backward from the R wave. The backward or reverse image sequence is intended to correct for fall-off in counts seen in the latter images when there are many arrhythmias. Images constructed from reverse gating would then have fewer counts in early systolic images. The two image sequences are joined at some point to create a single image sequence. Typically, the first two-thirds of the forward image sequence and the last one-third of the backward image sequence are joined. This technique is useful for analysis of the ventricular filling phase of the cardiac cycle.

The disadvantages of list mode MUGA are the great amount of disk space required for the data collection and the data processing time. A frame mode MUGA requires significantly less space on disk. For example, 24 frames of the cardiac cycle in single-view MUGA acquired in a 64 × 64-word mode matrix would require approximately 98 kilobytes of disk space.

First-Pass Study

The first-pass study is also referred to as first-pass radionuclide angiography. This method of dynamic cardiac imaging involves recording the first transit of a bolus (small volume) of radioactivity through the great vessels, pulmonary circulation, and the heart. The injection technique is critical to a quality study. A good bolus injection gives a distinct sequential visualization of the superior vena cava, right atrium, right ventricle, pulmonary circulation, left atrium, left ventricle, and the aorta (Fig. 15.3).

First-pass studies can be performed with the patient either supine or upright. A first-pass study is usually performed for right ventricle EF determination. A right anterior oblique position is optimal for a right ventricle EF study. A left anterior oblique position is optimal for left ventricle EF determination. The detector is placed as close to the chest as possible. Serial studies can be performed on the same day if the radiopharmaceutical used is quickly cleared from the blood.

No less than 20 frames per second must be acquired to assess heart function. Today, state-of-the-art computers are capable of acquiring and storing up to 100 frames per minute with count rates up to 200,000 counts/second, making list mode first-pass acquisition obsolete.

The total acquisition time is usually around 30 seconds. The bolus will pass through the right side of the heart more quickly (4–8 seconds) than through the left side of the heart (6–12 seconds). This difference is a result of the dilution of the bolus in the pulmonary circulation.

Comparison of Methods: Multiple Gated and First-Pass Acquisitions

Multiple Gated Acquisition

Images of the cardiac cycle obtained by MUGA have a high count density and therefore good resolution. This is the major advantage of the MUGA method for left ventricle EF and wall motion assessment. Obtaining a MUGA study in several different views is necessary to adequately assess wall motion. Regional wall motion and left ventricular function also can be evaluated from a MUGA study.

Acquisition of MUGA takes longer than the first-pass technique. MUGA can be performed up to 12 hours after the radiopharmaceutical has been administered.

The radiopharmaceutical equilibrates in the blood stream before acquisition. Therefore, it is not possible to view the right ventricle without some overlapping activity from the left side and/or great vessels. For this reason, right ventricle EF is best evaluated with a first-pass study.

First-Pass Method

A first-pass study provides data from only several cardiac cycles resulting in a lower count density and resolution than MUGA studies. The transit of the bolus through the right and left sides of the heart can be viewed and evaluated separately. With proper acquisition and processing techniques, reliable EF measurements can be obtained for both the right and left ventricles.

References

1. Reiber JH. Quantitative analysis of left ventricular function from equilibrium gated blood pool scintigrams: an overview of computer methods. *Eur J Nucl Med.* 1985;10:97–110.
2. Wallis, JW, Wu-Connolly L, Rocchini AP, Juni E. Dynamic arrhythmia filtration for gated blood pool imaging: validation against list mode technique. *J Nucl Med.* 1986;1347–1352.
3. Corbeitt JR, Jansen DE, Willerson JT. Radionuclide ventriculography: I. Technical aspects. *Am J Physiol Imaging.* 1987;2:33–43.
4. Bacharach SL, Green MV, Borer JS. Instrumentation and data processing in cardiovascular nuclear medicine: evaluation of ventricular function. *Semin Nucl Med.* 1979;9:257–274.
5. Watson NE Jr, Cowan RJ, Ball JD. Conventional radionuclide imaging. *Radiol Clin North Am.* 1994;32:477–500.
6. Berger HJ, Matthay RA, Pytlik LM, Gottschalk A, Zaret BL. First pass radionuclide assessment of right and left ventricular performance in patients with cardiac and pulmonary disease. *Semin Nucl Med.* 1979;9: 55–75.
7. Lee KH. *Computers in Nuclear Medicine. A Practical Approach.* New York, NY: Society of Nuclear Medicine; 1991:145–188.

Chapter 16

Red Blood Cell Labeling

The radiopharmaceutical routinely used for gated equilibrium blood pool imaging is ^{99m}Tc-pertechnetate (TcO_4) tagged to red blood cells (RBCs). Normally, ^{99m}Tc-TcO_4 diffuses passively into RBCs, forming an equilibrium inside and outside the cells. RBCs need to be "pretreated" with stannous ion (tin; Sn 2+) before labeling with ^{99m}Tc-TcO_4. Stannous ions are taken up by the RBCs. The tin reduces the ^{99m}Tc-TcO_4 inside the cells so that it remains intracellularly in the cytoplasm, on the cell membrane, or bound to hemoglobin (*1,2,3*).

Stannous Pyrophosphate Preparation and Supply

Cold stannous pyrophosphate (tin) is commonly used to pretreat the RBCs for labeling. The amount of tin used is critical to the procedure. The optimal dose of tin is 0.5–1.0 mg or approximately 15 vg/kg of body weight. If too little tin is used, the dose of ^{99m}Tc-TcO_4 will not completely label the RBCs. If more tin is used than is needed to sensitize all the RBCs, some tin will circulate freely and reduce ^{99m}Tc-TcO_4 outside the RBCs. In either case, the study will have an increased level of background activity as a result of free pertechnetate.

The incubation period for the tin and the RBCs must begin 15–30 minutes before the addition of the ^{99m}Tc-TcO_4 for adequate labeling. RBCs may be labeled with stannous pyrophosphate by three different methods. Each method has its advantages and disadvantages. The labeling efficiency varies for each method. A number of drugs, medical conditions, and technical factors can affect the efficiency of RBC labeling.

Stannous pyrophosphate is supplied in 1–3-mg vials that must be reconstituted with saline. The manufacturer's recommendations for stannous ion kit storage and reconstitution should be adhered to for optimal labeling results.

Methods

The three techniques for labeling RBCs are the in vivo, in vitro, and modified in vivo methods. The procedure for RBC labeling with each method and labeling efficiency are given in Table 16.1. The higher the labeling efficiency, the lower the background and the better the resolution of the ventricles when performing multiple gated acquisition

TABLE 16.1

Methods for Labeling Red Blood Cells with ^{99m}Tc-Pertechnetate (^{99m}Tc-TcO_4)

In vivo method (labeling efficiency 85%)

1. Inject (preferably by direct injection into a vein) appropriate amount of cold stannous pyrophosphate.
2. Wait 20–30 minutes for the stannous pyrophosphate to circulate and equilibrate.
3. In a different vein (preferably on the other arm), inject 20–30 mCi ^{99m}Tc-TcO_4.
4. Wait 5–10 minutes for the red blood cells (RBCs) to label and equilibrate within the body before imaging.

In vitro method (labeling efficiency 95%)

1. Using an 18–19-gauge needle, draw 10–12 mL of patient's blood into a syringe containing an appropriate anticoagulant, such as acid citrate dextrose, and the dose of tin.
2. Mix the syringe contents gently and allow to incubate at room temperature for 20 minutes.
3. Transfer the contents of the syringe to a sterile vial and centrifuge. The supernatant is discarded.
4. Add 4–6 mL saline to the cells, mix gently, and recentrifuge to remove all traces of excess tin.
5. Add 20–30 mCi ^{99m}Tc-TcO_4 to the blood cells and allow to incubate for 10 minutes before reinjection into the patient. This step requires appropriate radiation shielding.

Modified in vivo labeling method (labeling efficiency ≃ 90%)

1. Inject the appropriate amount of cold stannous pyrophosphate.
2. Wait 20–30 minutes for the stannous pyrophosphate to circulate and equilibrate.
3. Using an 18–19-gauge needle, draw 3–5 mL of the patient's blood into a shielded syringe containing a small amount of appropriate anticoagulant and 20–30 mCi ^{99m}Tc-TcO_4
4. Incubate RBCs and ^{99m}Tc-TcO_4 for 10 minutes, with occasional gentle mixing before reinjecting into the patient.

(MUGA). The labeling efficiencies listed in the table are those expected under optimal labeling conditions. Ejection fraction and wall motion can be assessed adequately when the optimal labeling efficiency is achieved for any of the methods listed (*4*).

In Vivo Method

The in vivo labeling method is the most commonly used for MUGA. Although the in vivo method yields the lowest labeling efficiency, there are some advantages to this method. It is the easiest method for the technologist, with no handling or manipulation of blood products or risk of inadvertent cross-transfusion. The small volume needed for a bolus injection with a first-pass study is difficult if not impossible to achieve with the other labeling methods. The in vivo method accommodates the performance of a first-pass study before MUGA, because the volume of the radiopharmaceutical dose can be equal to or less than 0.4 mL.

Modified In Vivo Method

In the modified in vivo labeling method, the tin-sensitized RBCs are incubated with $^{99m}Tc\text{-}TcO_4$ in a syringe. The addition of this one step improves labeling efficiency with a minimal amount of blood handling.

In Vitro Method

The in vitro technique yields the highest labeling efficiency (*4–7*). However, the technique is more difficult and time consuming. Both the in vitro and modified in vivo methods require specific safety policies and procedures to ensure against a cross-transfusion misadministration.

UltraTag Kit

In the in vitro method, RBCs and tin are incubated in a syringe or test tube. Before the addition of $^{99m}Tc\text{-}TcO_4$, the free tin must be removed from the blood by centrifuging and washing. In the in vitro method, the steps in the procedure for removing free tin can be avoided by adding sodium hypochlorite to oxidize extracellular tin before adding $^{99m}Tc\text{-}TcO_4$ to the blood mixture. This technique retains the high labeling efficiency of the in vitro method (*8*). The UltraTag RBC Kit (Mallinckrodt, Inc., St. Louis, MO) is a commercially available kit for this method of in vitro labeling. The kit requires only 1–3 mL of the patient's blood for labeling. A modification of the standard kit preparation procedure has been suggested for first-pass study doses (*9*).

Factors Affecting Red Blood Cell Labeling

Factors that can affect RBC labeling are listed in Table 16.2. The effects of too little or too much stannous ion have been discussed previously in this chapter. Pyrophosphate should not be injected through a heparin trap. Heparin may cause a reduction of the stannous ion, thereby affecting the labeling efficiency (*3*).

TABLE 16.2

Factors that Negatively Affect Red Blood Cell Labeling

Technical factors
Excess stannous ion: causes $^{99m}Tc\text{-}TcO_4$ reduced outside the red blood cells (RBCs).
Insufficient stannous ion: causes incomplete reduction of $^{99m}Tc\text{-}TcO_4$ inside the RBCs.
Prolonged generator in-growth: causes increased carrier ^{99}Tc.
Drugs such as heparin (noncirculating) or dextrose: will complex with $^{99m}Tc\text{-}TcO_4$
Other drugs
Cardiovascular:
Hydralazine
Prazosin
Propranolol
Methyldopa
Ca^{2+} blockers
Digoxin
Quinidine
H_2 blockers
Antibiotics:
Penicillin
Gentamycin
Metronidazole
Anti-inflammatory agents:
Acetylsalicylic acid and derivatives
Tranquilizers:
Benzodiazepines
Chemotherapy agents:
Doxorubicin
Gastrointestinal:
Ranitidine
Other factors
Decreased hematocrit
Iodinated contrast agent administered within 24 hours of labeling
Immune disorders

$^{99m}Tc\text{-}TcO_4$ should not be injected through intravenous lines containing substances with which it is known to complex, such as dextrose and heparin. The formation of a ^{99m}Tc complex will reduce the amount of $^{99m}Tc\text{-}TcO_4$ available for labeling to RBCs, lower the labeling efficiency, and increase the level of background in a study. For example, ^{99m}Tc-labeled heparin is primarily taken up in the liver and kidneys (*10*). A therapeutic dose of heparin circulating in the blood stream is not problematic, and RBC labeling can be accomplished adequately. For patients receiving heparin, a slightly higher dose of stannous ion (20 mg/kg of body weight) will improve labeling efficiency (*11*).

A number of other drugs have been reported in the literature to interfere with RBC labeling. Iodinated contrast agents administered within 24 hours of RBC labeling have been reported to decrease labeling efficiency, although the exact cause is not known (*12*).

An excess of nonradioactive or carrier ^{99}Tc will lower RBC labeling efficiency by competing with the radioactive form. The ^{99m}Tc-TcO_4 used for RBC labeling should come from a generator elution that was preceded by another generator elution not more than 24 hours before. Generator in-growth times of greater than 48 hours will significantly affect labeling efficiency (*13*).

References

1. Hambye AS, Vandermeiren R, Vervaet A, Vandevivire J. Failure to label red blood cells adequately in daily practice using an in vivo method: methodological and clinical considerations. *Eur J Nucl Med.* 1995;22:61–67.
2. Linder KE, Nunn AD, Strauss HW. Radiopharmaceuticals for cardiac imaging. In: Sandler MP, Patton JA, Coleman RE, et al. *Diagnostic Nuclear Medicine.* Vol. 1. 3rd ed. Baltimore, MD: Williams & Wilkins; 1996:217–229.
3. Murphy PB, Port SC. Radionuclide evaluation of left ventricular function. In: Sandler MP, Patton JA, Coleman RE, et al. *Diagnostic Nuclear Medicine.* Vol. 1. 3rd ed. Baltimore, MD: Williams & Wilkins; 1996:403–442.
4. Hegge FN, Hamilton GW, Larson SM, Ritchie JL, Richards P. Cardiac chamber imaging: a comparison of red blood cells labeled with Tc-99m in vitro and in vivo. *J Nucl Med.* 1978;19:129–134.
5. Neumann P, Schicha H, Schurnbrand P, Bahre M, Emrich D. Visualizing cardiac blood pool: comparison of three labeling methods. *Eur J Nucl Med.* 1983;8:463–466.
6. Saha GB. *Fundamentals of Nuclear Pharmacy.* 3rd ed. New York, NY: Springer-Verlag; 1992:109–142.
7. Callahan RJ, Froelich JW, McKusick KA, Leppo J, Strauss HW. A modified method for the in vivo labeling of red blood cells with Tc-99m: concise communication. *J Nucl Med.* 1982;23:315–318.
8. Patrick ST, Glowniak JV, Turner FE, Robbins MS, Wolfangel RG. Comparison of in vitro RBC labeling with the Ultra Tag RBC kit versus in vivo labeling. *J Nucl Med.* 1991;32:242–244.
9. Stallings LE, Cheng KT, Spicer KM, Wiegner R, Humphries JK. Ultra Tag RBC kit for combined cardiac first-pass and multigated acquisition studies. *J Nucl Med Technol.* 1997;25:44–48.
10. Porter WC, Dees SM, Freitas JE, Dworkin HJ. Acid-citrate-dextrose compared with heparin in the preparation of in vivo/in vitro technetium-99m red blood cells. *J Nucl Med.* 1983;24:383–387.
11. Rao SA, Knobel J, Collier BD, Isitman AT. Effect of Sn(II) ion concentration and heparin on technetium-99m red blood cell labeling. *J Nucl Med.* 1986;27:1202–1206.
12. Tatum JL, Burke TS, Hirsch JI, Miller WW, Fratkin MJ. Pitfall to modified in vivo method of technetium-99m red blood cell labeling. Iodinated contrast media. *Clin Nucl Med.* 1983;8:585–587.
13. Kelly MJ, Cowie AR, Antonio A, Barton H, Kalff V. An assessment of factors which influence the effectiveness of the modified in vivo technetium-99m-erythrocyte labeling technique in clinical use. *J Nucl Med.* 1992;33:2222–2225.

Chapter 17

Resting Gated Acquisition

This chapter will cover the technical aspects of resting gated acquisition. Pitfalls and problems associated with acquisition that can reduce the diagnostic quality of the study will be discussed.

Electrocardiograph Placement

For resting multiple gated acquisition (MUGA), the patient must be connected to an electrocardiographic (ECG) gating device that is also interfaced with the computer. Three leads adequately provide the basic signal for gating. These leads are:

- The right arm lead, which is placed in the region of the right shoulder or the right clavicle;
- The left arm lead, which is placed in the region of the left shoulder or the left clavicle; and
- The left leg lead, which is placed on the abdomen just below the ribs. Metal snaps and wires must be kept out of the field of view.

It is critical that the signal from the leads be free of interference. This requires proper skin preparation before lead placement. To remove dead skin and surface oils, the skin should be rubbed and cleansed with an abrasive pad saturated with alcohol. Hair should be shaved, if necessary.

Pre-jellied electrodes with a foam backing are used routinely. The surface center that will have contact with the skin must be moist for good conduction. The electrode should be pushed firmly to the skin. Electrodes should not be placed on scar tissue, which does not give good conductivity.

Common Problems with the Electrocardiograph Connection

The goal is to provide the computer a signal with a strong positive R wave. The R wave triggers acquisition. Occasionally the placement of the leads on the chest will need to be adjusted to produce a single physiological trigger. A list of common problems encountered with the ECG signal and appropriate remedies are listed in Table 17.1.

Views/Patient Positioning

To assess wall motion, two or three views of the heart are needed. These are the anterior, best septal, and 70° left anterior oblique (LAO) or lateral views. The camera is positioned as close to the chest as possible for all views. The patient is positioned supine on the imaging table for the anterior and best septal views. The left arm must be raised over the head to get the camera as close as possible to the chest for the best septal view. For all views, patient comfort is essential to minimize motion and heart rate irregularity.

Best Septal View

The best septal view is an LAO view that gives the best separation of the right and left ventricles of the heart. This view is used to calculate the left ventricular ejection fraction (EF). The exact angle will vary from patient to patient, and care must be taken to choose it appropriately. The best septal view is usually in the range of 30°–50° LAO.

It is critical that the left ventricle be isolated from all other structures in the chest in this view. A small amount (approximately 5°–10°) of caudal tilt is required to sepa-

TABLE 17.1

Common Problems Associated with the Electrocardiographic Signal for Gating

Problem	Cause	Remedy
Inverted R wave	Reversed polarity	Criss-cross the leads
Tall T wave	May cause double gating	Move leads around, bring leg lead closer to the heart
Pacemaker spike	May cause double gating	
Bad tracing	Poor electrode contact with the skin and/or wire fractures.	Remove the pads, clean the skin, and reapply new pads. Check the wires often for cracks and breaks and replace as needed.

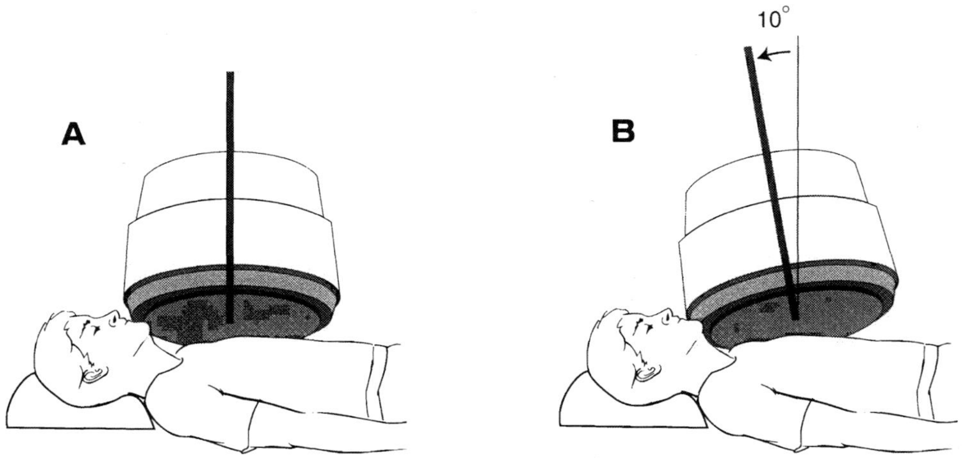

Figure 17.1 Caudal tilt of the camera. In both pictures the camera is angled 45°. (A) No caudal tilt. (B) The camera is angled with a 10° caudal tilt. It is tilted back toward the head of the imaging table at approximately 10°.

rate the left ventricle from the left atrium. Caudal tilt is achieved by angling the camera toward the patient's head (Fig. 17.1). In the best septal view, the left ventricle looks like an upright oval in diastole (Fig. 17.2A).

Anterior View

The anterior view is positioned 45° anterior to the best septal view. No caudal tilt is used for this view.

Lateral View

The left lateral view is 45° higher than the best septal view. For the lateral view, the patient is positioned on his or her right side and the camera is placed directly against the left lateral chest wall. Again, caudal tilt is not required. If the patient is not able to assume the decubitus position, a cross-table lateral view can be acquired with the patient supine. In either case, the patient's arms are raised out of the field of view.

Right Ventricle View

A right ventricular EF is routinely determined from a first-pass study. MUGA is less than optimal for right ventricular EF determination because of interfering activity from the adjacent structures, including other cardiac chambers and great vessels. If a right ventricular EF is to be determined from a MUGA, this special view is used (Fig. 17.3). The right ventricle view is 20° anterior to the best septal view. Five degrees of caudal tilt are required for this view.

Camera and Computer Set-Up

A 20% energy window and a low-energy, all-purpose or high-resolution collimator are used for acquisition. The heart should occupy 50% of the field of view. When using a large-field-of-view camera, this will require some acquisition zoom to reduce the field of view and exclude counts from other structures, such as the liver and spleen.

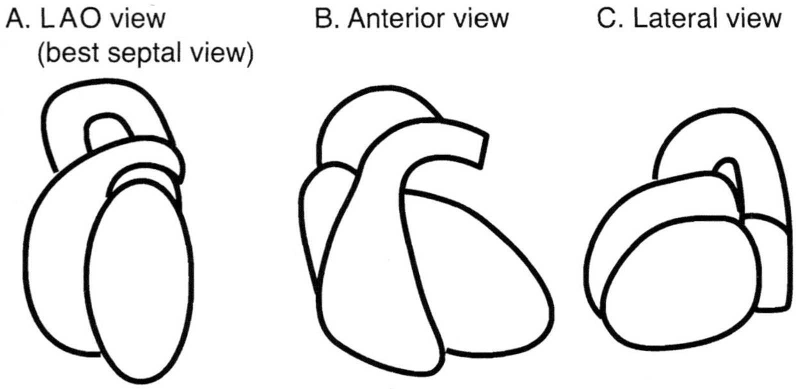

Figure 17.2 Proper positioning of commonly used views for multiple gated acquisition. LAO = left anterior oblique.

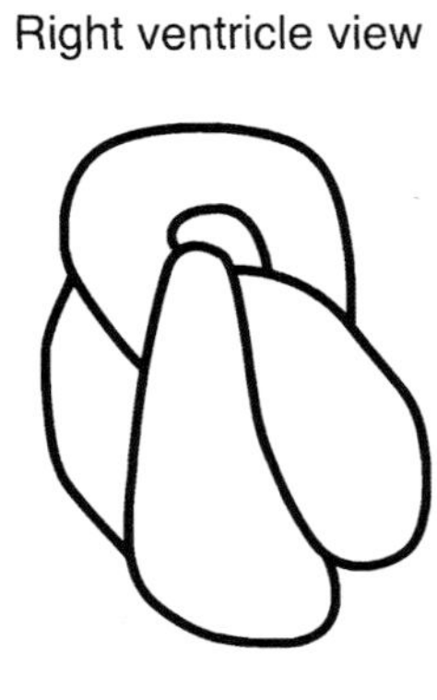

Figure 17.3 Right ventricle view.

TABLE 17.2

Optimal Statistics for Resting Multiple Gated Acquisition Study

(Best Separation Left Anterior Oblique View)

Use any of the following:	
Total study counts (24 frames/cardiac cycle)	6–8 million
Minimum counts/frame	250,000
Counts/pixel over the left ventricle	250

If diastolic parameters (such as peak filling rate and time to peak filling) are to be evaluated, a narrow R-R interval window is needed to eliminate bad beats (see chapter 15). The optimal acquisition is accomplished with an R-R interval window of 10%–15%. When the heart rate is irregular, a wider R-R interval window can be used.

For the best septal view, a minimum of 24 frames of the cardiac cycles is preferred. The higher the number of frames, the better is the quality of the study for assessing EF. The systolic and diastolic contours of the heart are better defined with a higher number of frames. A minimum of 16 frames of the cardiac cycle is adequate for other views intended for wall motion review.

MUGA is a high-resolution study. The field of view for the heart is small. These factors affect the selection of the matrix type and size. A 32 × 32 word mode matrix may be adequate. A 64 × 64 word mode matrix is most often used.

Method of Acquisition Termination

The greater the number of frames of the cardiac cycle, the better will be the temporal resolution of the study. A minimum number of 250,000 counts per frame is required for statistical reliability of the EF and image quality. As the number of frames increases, it will take longer to acquire the desired counts per frame.

Before beginning a MUGA study, a method of termination must be defined. There are four possible methods of termination: the counts/frame, total study counts, the number of cardiac cycles, and total time. Any method is acceptable, as long as an adequate number of counts is acquired. Adequate statistics are required for a good quality study. Table 17.2 lists the optimal counts needed for the best septal LAO view, which will be used for EF calculations. Half as many counts are adequate for the other views.

Dose and Procedure

The radiopharmaceutical of choice for MUGA is ^{99m}Tc-TcO_4 tagged to red blood cells. Labeling methods are described in chapter 16. After the radiopharmaceutical is administered, the labeled red blood cells must equilibrate in the body. This takes a minimum of 5 minutes. The patient can be connected to the ECG and positioned under the camera during this waiting period. The patient should be instructed to relax and not move during acquisition. Excessive conversation can cause an irregular heart rhythm and should be avoided after acquisition has started. The technologist should monitor the ECG tracing throughout the duration of the acquisition, which generally takes 5–10 minutes per view.

References

1. Wittry MD, Juni JE, Royal HD, Heller GV, Port SC. Procedure guidelines for equilibrium radionuclide ventriculography. Society of Nuclear Medicine. *J Nucl Med.* 1997;38:1658–1661.

2. Garcia EV. Imaging guidelines for nuclear cardiology procedures. Part I. *J Nucl Cardiol.* 1996;3:G3–G46.

3. Klingersmith, WC, Eshima D, Goddard J, eds. *Nuclear Medicine Procedure Manual. 2000–2002.* Englewood, CO: Wick Publishing; 1999–2000.

Chapter 18

Data Processing of Gated Studies

To create a better visual display for wall motion analysis, the frames of the cardiac cycle acquired by multiple gated acquisition (MUGA) are processed to correct for statistical variance and the normal variation that occurs even with a regular heart rate. The data processing includes some spatial and temporal smoothing and frame normalization (*1*). These processing steps are performed automatically by the computer before cine display.

Initial Data Processing and Edge Detection

A patient's left ventricular ejection fraction (EF) and other quantitative information are obtained from the best separation left anterior oblique view. Quantitation begins with a determination of the net counts in the region of the left ventricle for each frame in the study. A precise edge around the perimeter of the left ventricle is crucial for accurate quantitation. Fully automatic and semiautomatic computer programs are available for left ventricle edge detection. These programs work extremely well when the there is a good target-to-background ratio. The target-to-background ratio is dependent on good positioning and red blood cell labeling efficiency. The edge around the ventricle also can be drawn manually, although this is not recommended. Even when the target-to-background ratio is optimal, manual region of interest (ROI) selection is very subjective and dependent on the way in which the eye visualizes a border or edge. Computerized edge detection is more objective, because it is based on a mathematical model and not on visual discrimination. Manually drawing an edge around the left ventricle in each frame of the study is also labor intensive. Semiautomatic and automatic edge detection software uses threshold and second derivative methods for finding the edge of the ventricle in each frame.

Automatic Edge Detection

The first step in automatic edge detection is finding the center of the left ventricle. Fully automated software can find the center of the left ventricle. The technologist may need to participate in adjusting the placement of the center of the ventricle or may be required to draw a box around the left ventricle in the first frame. From the center of the left ventricle, numerous radii are sent outward in all directions to search for the edge using the thresholding and second derivative methods. The threshold value is a predetermined percentage of the average pixel count in the center of the ventricle. The computer draws edges based on the threshold values. The computer starts with a high threshold value and gradually lowers it. As the threshold value is lowered, the edge of the ventricle moves outward from the center along each radius (Fig. 18.1). As the edge moves outward from the center along each radius or line, there will be a gradual pixel-to-pixel change in counts. At the true edge of the ventricle, there will be a sudden great change in counts compared with previous pixel-to-pixel changes. The second derivative of the line, which is this point of great change or the point of the steepest decline, is the true edge of the ventricle (*2–4*). The edge search along each radius or line is terminated when this point is reached.

The ROI generated for each frame of the cardiac cycle should include all of the left ventricle. The region should not include or overlap any other structures, such as part of

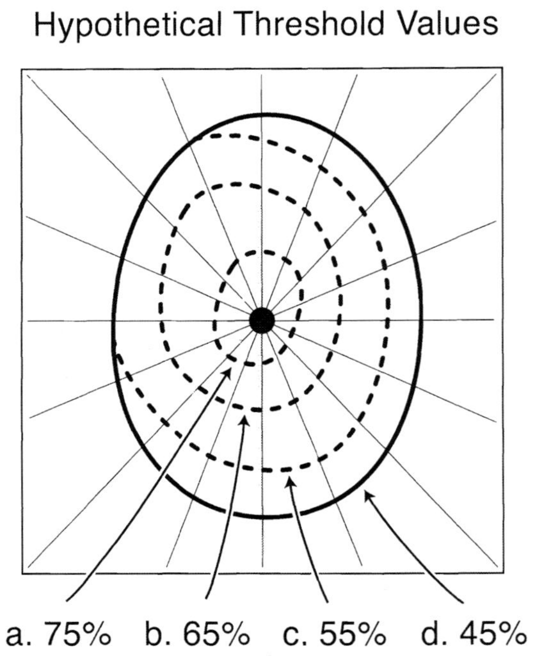

Figure 18.1 Example of automated edge detection routine–Threshold/Second derivative method. The dotted lines represent an edge corresponding to a specific threshold value. The solid line represents the threshold value at which the edge search is terminated.

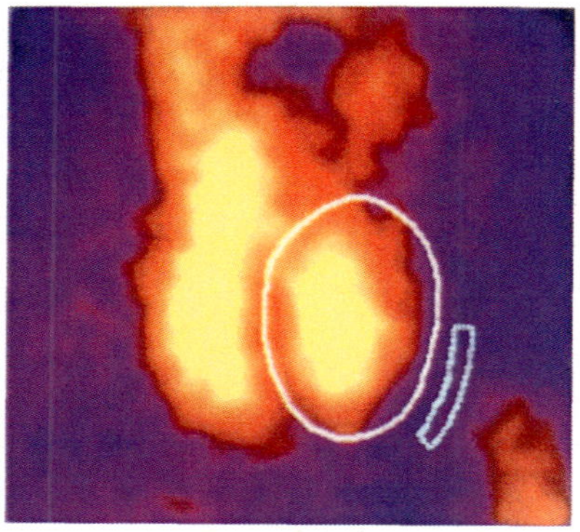

Figure 18.2 Placement of background region of interest.

the right ventricle, great vessels, or left atrium. Computer-generated ROIs should be inspected visually for irregular contours extending beyond the borders of the left ventricle. Occasionally, irregularities in the ventricular ROI occur. Irregular regions may be the result of any of the following: the presence of an aneurysm, an enlarged heart, poor positioning so that a good separation is not achieved between the left ventricle and left atrium or right ventricle, vascular anomaly, a high background from poor red blood cell labeling, obesity, or other causes. Specific attention should be paid to the regions drawn around the end diastolic (ED) and end systolic (ES) frames. Computer-generated irregular regions should be corrected manually. If correction to the ROIs is not made as needed, the calculated EF and other parameters associated with the volume curve will not be accurate.

TABLE 18.1

Data and Formulas for Ejection Fraction Calculation

Data

Total counts in end diastole (ED) region of interest (ROI)

Number of pixels in ED ROI

Total counts in end systolic (ES) ROI

Number of pixels in ES ROI

Average counts/pixel in background ROI

Formulas

$$\text{Ejection fraction} = \frac{\text{net ED counts} - \text{net ES counts}}{\text{net ED counts}} \times 100.$$

Net ED counts = total counts in ED ROI − (number of pixels in ED ROI × average counts/pixel in background ROI).

Net ES counts = total counts in ES ROI − (number of pixels in ES ROI × average counts/pixel in background ROI).

TABLE 18.2

Reasons for Errors in Ejection Fraction Measurements

Overestimation

Too much background subtraction (e.g., region of interest [ROI] over the spleen or aorta)

Post basal aneurysm status

Pulmonary hypertension

Underestimation

Including the left atrium in the left ventricle ROI in systole, especially when the left atrium is enlarged

Poor separation of the right and left ventricles

Too little background subtraction (e.g., left ventricular hypertrophy, pleural or pericardial effusion)

Variable gating

Presence of anterior aneurysm

Background Region of Interest

The numbers used to calculate EF and generate the volume curve should be the counts in each frame from the blood pool within the ventricle. For an accurate EF and volume curve, background activity must be subtracted from the total counts of each left ventricle ROI. Counts coming from the overlying structures, such as the chest wall, must be subtracted from the total region counts. To do this, a background ROI must be generated to determine the average background counts per pixel. A background ROI should be drawn along the lateral side of the left ventricle edge. The background ROI should be placed only several pixels away from the left ventricle edge so that it is close to but not touching the heart. It should be several pixels wide and horizontal but somewhat crescent shaped as it contours to the shape of the ventricle (Fig. 18.2). The background ROI approximates activity in structures overlying the ventricle blood pool. Automated edge detection software also generates the background ROI. The background ROI must be inspected to make sure that it is drawn appropriately and does not touch other structures. The background ROI should not touch or overlie the spleen, aorta, right or left ventricle, an aneurysm, or other blood pool structure.

Ejection Fraction Determination

The first or second frame of the study will have the highest counts in the left ventricle ROI, and that frame is the ED frame. The frame with the lowest number of counts in the left ventricle ROI is the ES frame. The data needed and formulas for calculating the EF are listed in Table 18.1. Many factors can cause an over- or underestimation of the EF (Table 18.2).

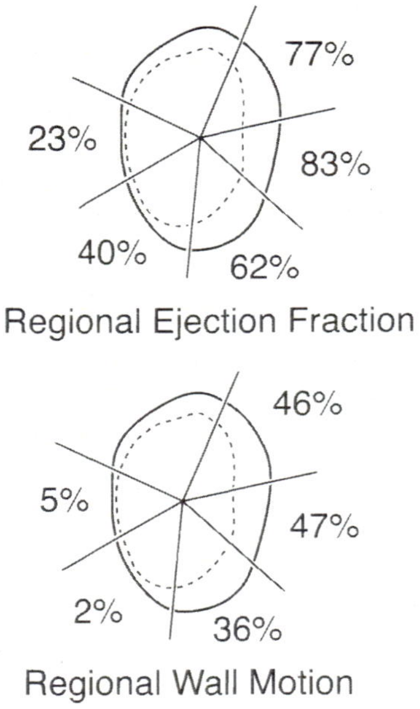

Figure 18.3 Regional ejection fraction.

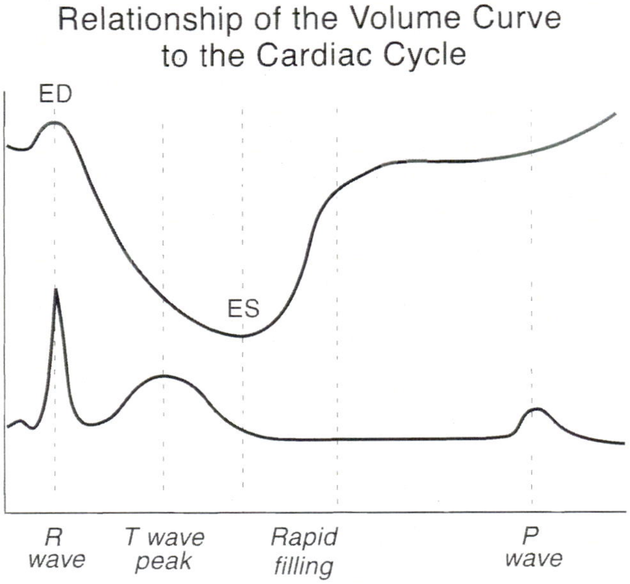

Figure 18.5 Relationship of the volume curve to the cardiac cycle.

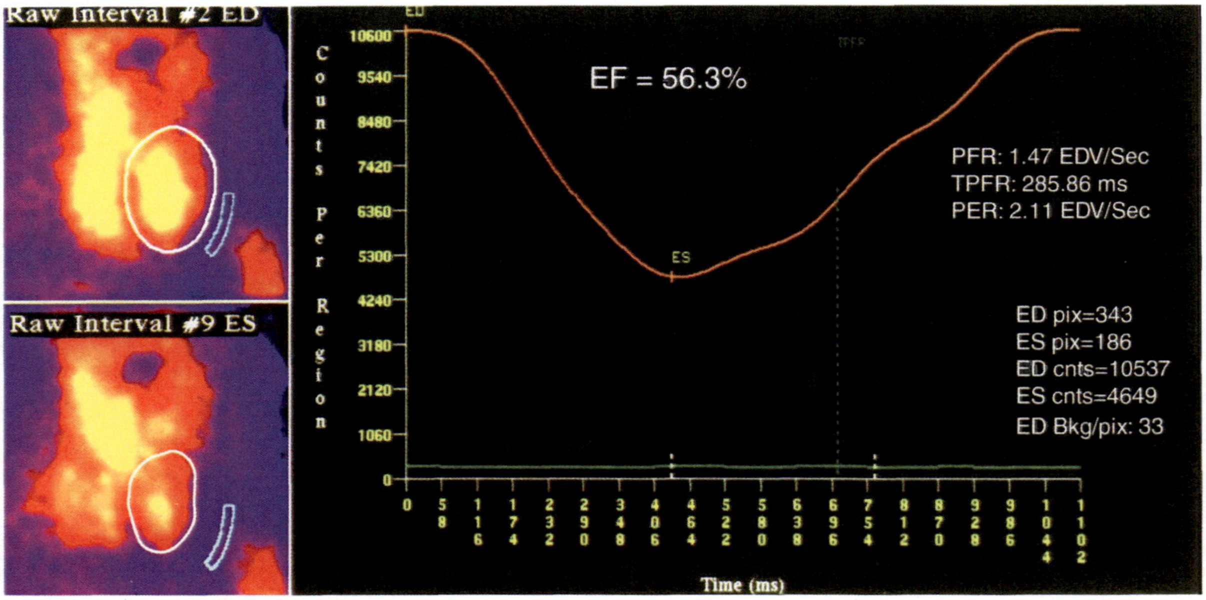

Figure 18.4 Normal images and volume curve.

Global and Regional Ejection Fraction

The global EF is calculated from the entire ventricular volume. The ventricle also can be divided into five to eight segments like a pie to evaluate regional EF. A volume curve is generated for each segment. The regional EF should vary from segment to segment. The septal side of the ventricle would normally have a slightly lower EF than the lateral wall or apex. The regional EF can be useful and can be correlated with the evaluation of regional wall motion (Fig. 18.3).

Generation of the Volume Curve and Associated Data

Figure 18.4 shows ED and ES images with their corresponding ROIs and volume curve for the determination of EF. A volume curve is generated by plotting the net counts of each left ventricle ROI as a function of time (5). The volume curve shows how the blood volume changes through the average cardiac cycle. Figure 18.5 shows how the volume curve relates to the normal cardiac cycle. As previously described, the peak of the R wave corresponds

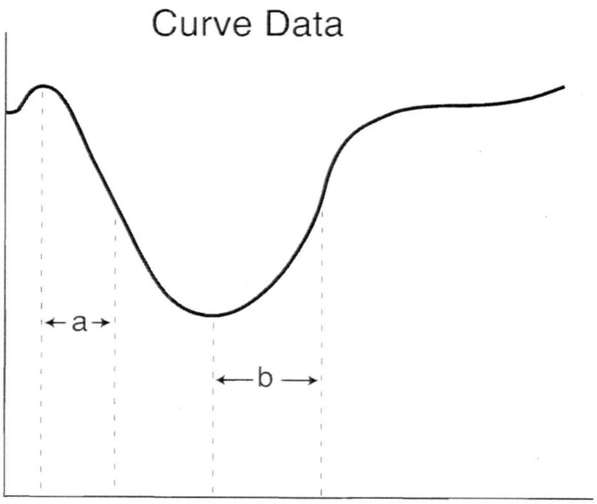

Figure 18.6 Curve data. (A) Time to peak emptying rate. (B) Time to peak filling rate.

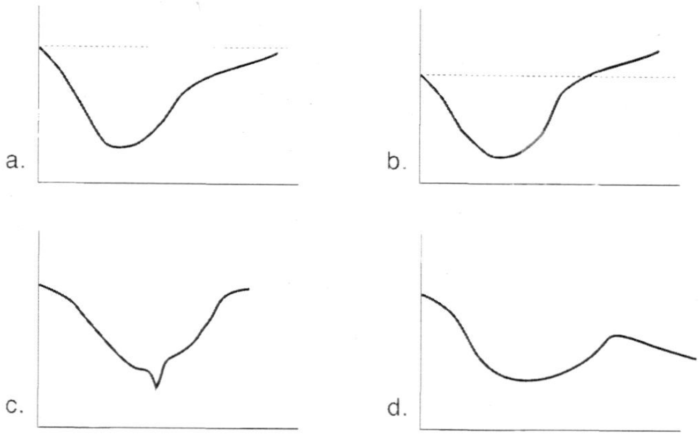

Figure 18.7 Diagrams of abnormal curves and associated causes. (A) Normal curve. (B) Reversing polarity causes the frames at the beginning of a study to have fewer counts than they should. The calculated ejection fraction (EF) will be falsely low. (C) Poor region of interest (ROI) selection on one frame. A disfigured curve can result from an improperly drawn ROI. The distortion of this curve will cause a falsely elevated EF. (D) Frame drop-off. Excessive arrhythmias will cause a loss of counts in the later frames of the study.

to ED. ES corresponds to the end of the T wave. The peak of the T wave corresponds to the end of the rapid filling phase of systole. Most ventricular filling occurs in a relatively short period of time, followed by a slower filling period. Atrial contraction begins with the P wave (*6*). Other data can be derived from the volume curve. Systolic parameters that can be derived from the volume curve include the peak emptying rate and time to peak emptying rate. Diastolic parameters include the peak filling rate and the time to peak filling rate (Fig. 18.6). Rates are obtained by measuring the slope of the curve and are expressed in ED volume per second (EDV/sec) (*7*).

Curves with abnormal shapes and the associated causes are shown in Figure 18.7. Figure 18.8 is a volume curve from a study of a patient in atrial fibrillation.

Image Data

A number of images can be generated from MUGA data that can be useful in the technical evaluation and interpretation of the study. The ED and ES frames can be shown with or without the left ventricle ROI superimposed. A stroke volume picture can be generated by subtracting the ES frame from the ED frame. A paradox image, which is an approximation of atrial stroke volume, can be generated by subtracting the ED frame from the ES frame. All negative pixels are set to zero. An amplitude picture is color scaled to show the amount or force of contraction. Figure 18.9 shows various image data.

Phase Analysis

The cine display of MUGA data shows the wall motion of the heart during the average cardiac cycle. Phase analysis is a type of parametric imaging that produces a display of heart contraction. Each pixel in the series of frames of the cardiac cycle can be transformed by Fourier analysis into temporal phase waves. Heart contraction is rhythmic, in that blood flow in the heart increases when the heart is filling and decreases when the heart is contracting. Graphically, the time–activity curve looks like a sine wave. If a time–activity curve is generated for each pixel in the study, background pixels will produce a somewhat straight horizontal line. Pixels in the region of the heart will produce a sine wave. Heart contraction begins in the atria and spreads through the ventricles. For each pixel, the phase of the wave will differ depending on its location in the heart. The phase angle is an arbitrarily chosen point on the sine wave, such as the peak or ED point. All pixels that are in the same phase of contraction are displayed with the same color. A phase histogram shows the number of pixels that are contracting together with the same phase angle (*8,9*) (Fig. 18.10).

The dynamic phase analysis display is very useful for assessing conduction abnormalities and dyskinetic wall motion, which are frequently associated with ventricular aneurysms.

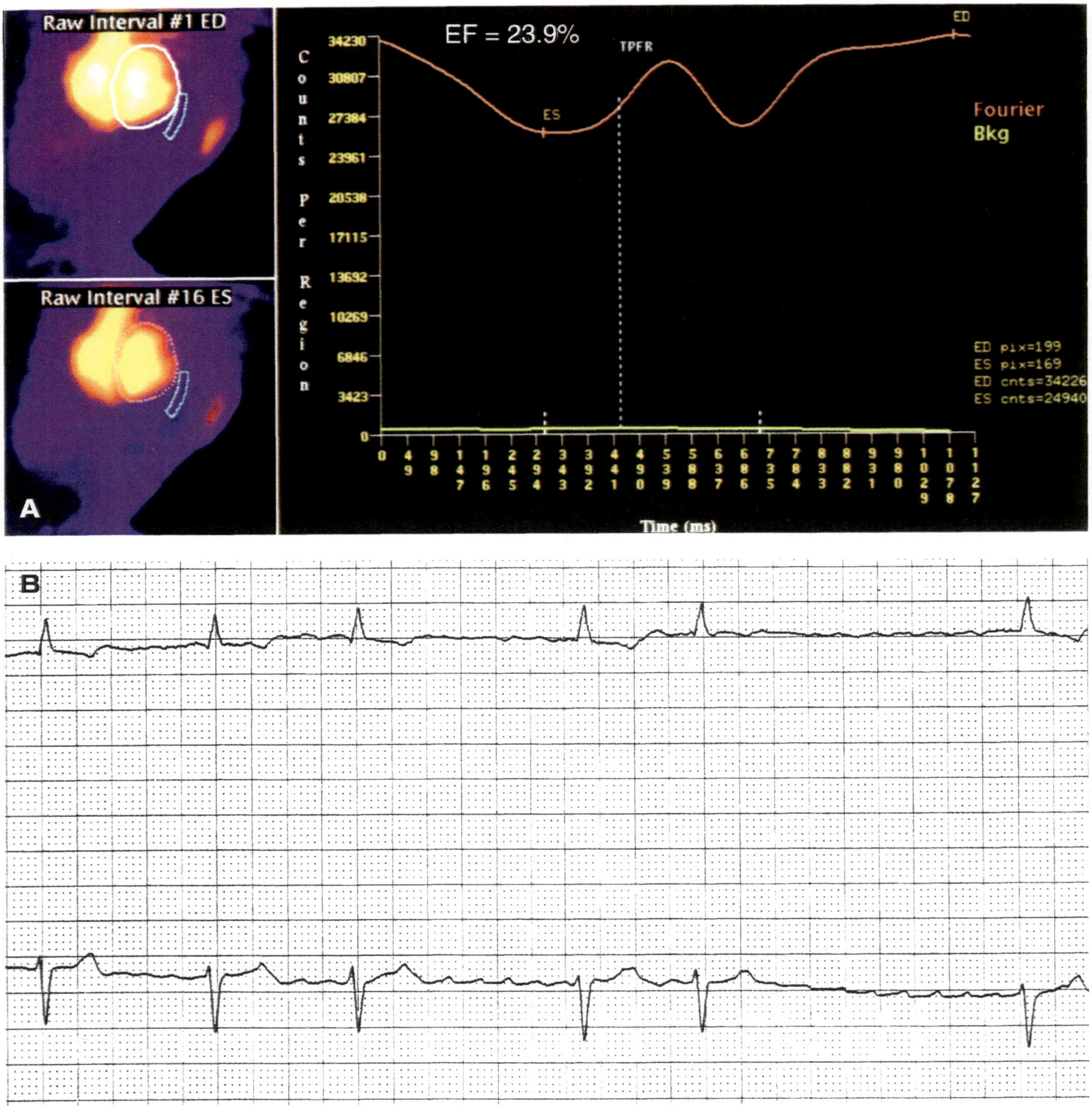

Figure 18.8 Volume curve of a patient with atrial fibrillation. (A) Curve. (B) Corresponding electrocardiograph.

References

1. Lee KH. *Computers in Nuclear Medicine. A Practical Approach.* New York, NY: The Society of Nuclear Medicine; 1991:147–188.
2. Lee KH. *Computers in Nuclear Medicine. A Practical Approach.* New York, NY: The Society of Nuclear Medicine; 1991:128–146.
3. Murphy PB, Port SC. Radionuclide evaluation of left ventricular function. In: Sandler MP, Patton JA, Coleman RE, et al., eds. *Diagnostic Nuclear Medicine.* 3rd ed. Baltimore, MD: Williams and Wilkins; 1996:403–442.
4. Chang W, Henkin RE, Hale DJ, Hall D. Methods for detection of left ventricle edges. *Semin Nucl Med.* 1980;10:39–53.

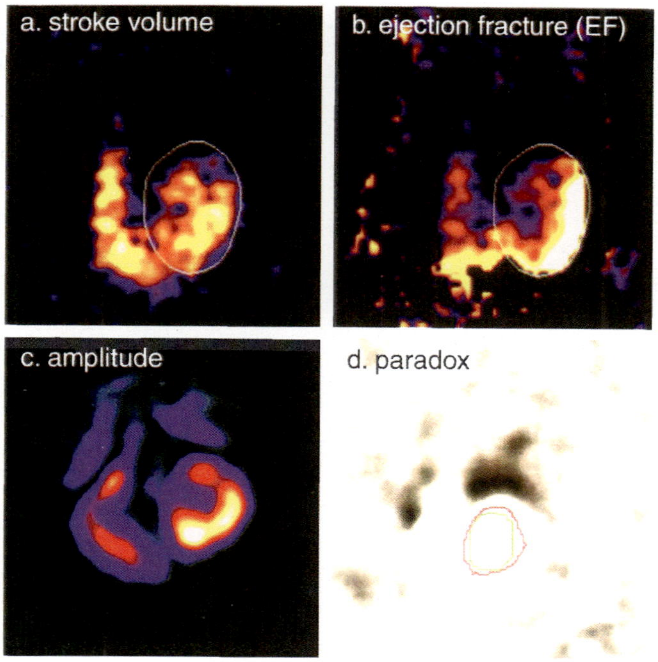

Figure 18.9 Examples of image data. (A) Stroke volume. (B) Ejection fraction. (C) Amplitude. (D) Paradox.

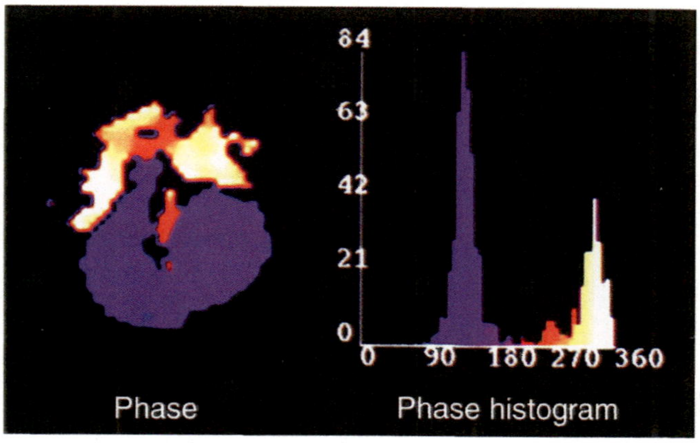

Figure 18.10 Phase analysis. Image and histogram.

5. Wagner RH. Sources of error in gated blood pool studies. *Appl Radiol.* 1991:41–44.
6. DePuey EG. Evaluation of cardiac function with radionuclides. In: Gottschalk A, Hoffer PB, Potchen EJ, eds. *Diagnostic Nuclear Medicine.* 2nd ed. Baltimore, MD: Williams & Wilkins; 1988:355–378
7. Garcia EV. Imaging guidelines for nuclear cardiology procedures. Part 1. *J Nucl Cardiol.* 1996;3:G3–G4.
8. Chang W, Thrall JH. Fourier phase and amplitude analysis of the radionuclide ventriculogram. Technique and application in functional cardiac imaging. *Software.* 1982:16–21.
9. Links JM, Douglass KH, Wagner HN Jr. Patterns of ventricular emptying by Fourier analysis of gated blood-pool studies. *J Nucl Med.* 1980;21:978–982.

Chapter 19

Exercise Gated Study

Cardiac blood pool imaging by multiple gated acquisition (MUGA) can be performed while a patient is exercising. The exercise gated study is useful for evaluating exercise-induced left ventricle dysfunction. An exercise gated study allows us to see the effects of ischemia on the wall motion and ejection fraction (EF) of the heart. The EF at peak exercise can also be calculated and compared with the EF at rest.

Exercise Gated Study Equipment

Upright treadmill exercise or bicycle exercise are the two optimal exercise methods, because exercise is most often conducted in the upright position. Several problems limit the use of treadmill or upright bicycle exercise with MUGA. It is difficult to position a patient's chest close to most gamma cameras and to limit motion when a patient is on a treadmill and/or stationary bicycle. For this reason, special bicycles and cardiac stress tables have been designed for performing the exercise gated study (Fig. 19.1). The exercise equipment (bicycle or stress table) is connected to an ergometer (Fig. 19.2). The ergometer is used to control and monitor the exercise workload throughout the study. The patient must maintain a constant number of revolutions per minute, which are monitored on the ergometer. The ergometer is also used to adjust the workload (measured in watts [W]) applied to the pedals.

Acquisition Procedure

The exercise study requires a series of acquisitions and is conducted in three stages: rest, exercise, and recovery. Before beginning exercise, a resting MUGA study is completed. The resting study is also called the baseline study. The patient then is asked to pedal at a constant number of revolutions per minute with a modest workload (25 W) applied to the pedals. A MUGA study is acquired for 2 minutes. The workload to the pedals is then increased by 25 W. The patient is given 10–15 seconds to adjust to the new workload before the second acquisition is started. The patient must maintain the same constant number of revolutions per minute. Again, acquisition lasts for 2 minutes. At the end of each acquisition, the workload to the pedals is increased by 25 W and another 2-minute acquisition is started.

Throughout the study, the patient's electrocardiograph (ECG), blood pressure, heart rate, and pulse are monitored. The patient is also asked to inform the staff of any chest discomfort, fatigue, or other symptoms of ischemia.

Acquisition is continued at increasing workload levels up to the patient's maximum stress capability. The exercise workload is increased every 2 minutes until symptoms of ischemia are precipitated (i.e., marked ECG changes, marked arrhythmia, or severe fatigue). For the study to be valid, stress must be adequate to precipitate ischemia if it exists. For the optimal study, the patient's heart rate should reach 85% of the maximum predicted heart rate. When

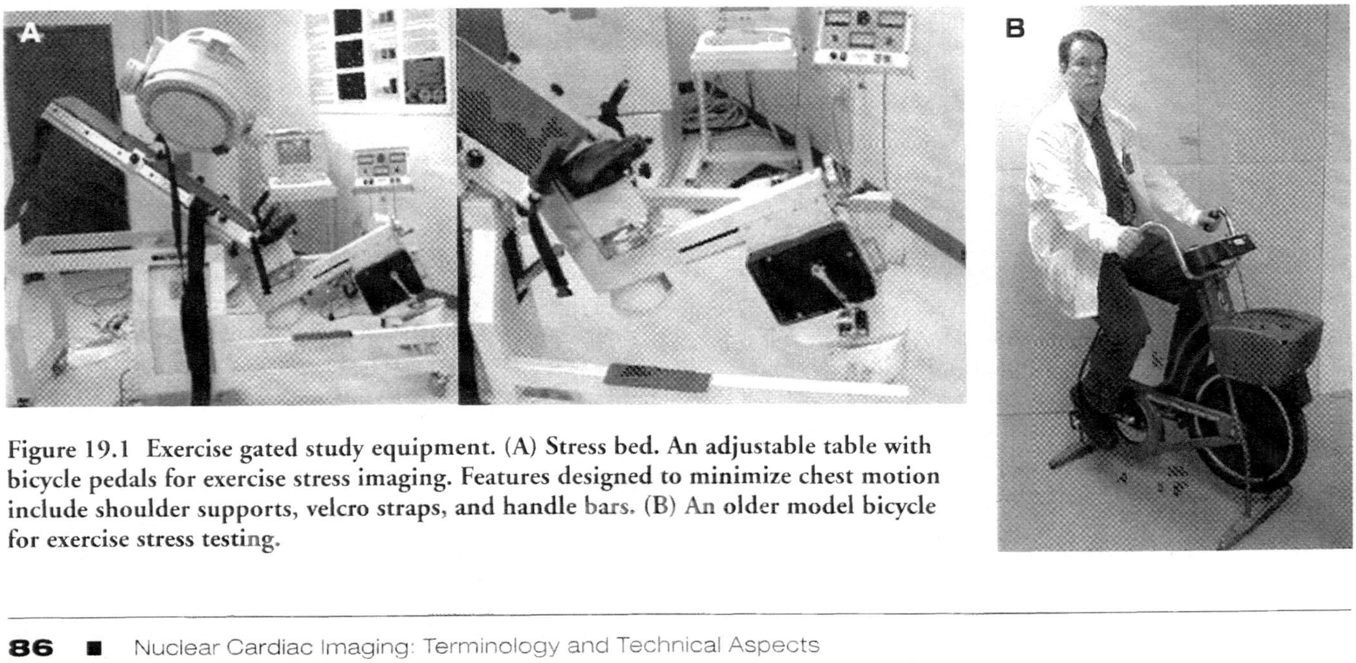

Figure 19.1 Exercise gated study equipment. (A) Stress bed. An adjustable table with bicycle pedals for exercise stress imaging. Features designed to minimize chest motion include shoulder supports, velcro straps, and handle bars. (B) An older model bicycle for exercise stress testing.

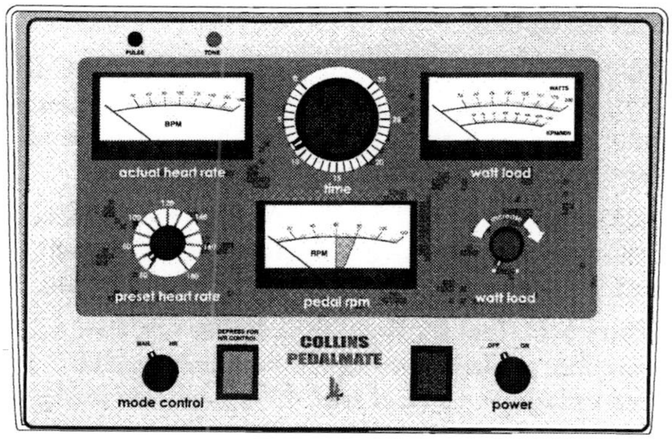

Figure 19.2 Ergometer.

his or her maximum exercise workload is reached, the patient will feel tired and want to stop pedaling. The patient should be encouraged to continue pedaling to complete the 2-minute acquisition at maximum stress, if possible. The final acquisition is made immediately after the maximum workload acquisition. This acquisition is referred to as the recovery study. The patient does not pedal during the recovery acquisition.

Leads for Gating

A 12-lead ECG connection is needed (see final section of this text lead placement). Some of the electrodes will be in the field of view. The electrodes used for this study must have plastic and not metal connectors.

Dose and Patient Preparation

Red blood cells are labeled with $^{99m}Tc\text{-}TcO_4$, as described in chapter 16. A slightly higher dose of $^{99m}Tc\text{-}TcO_4$ (25–30 mCi) is recommended to improve image statistics.

The patient is positioned for the best septal left anterior oblique (LAO) view and should be given an opportunity to practice pedaling to ensure that the equipment is adjusted comfortably for his or her height and to minimize motion. Excessive practice that raises the heart rate before acquisition should not be allowed.

The patient should fully understand what will be expected. A very good procedure explanation is essential.

Camera and Computer Set-Up

Stress MUGA acquisition is shorter in duration than resting MUGA acquisition. Although the best separation LAO view is used for stress MUGA, 16 frames of the cardiac cycle are recommended. Counts per frame are increased by reducing the number of frames from 24 to 16.

A high-sensitivity or low-energy, all-purpose collimator can be used. The R-R interval window can be increased to 25%, because an increase in arrhythmias is expected with exercise. The method of termination used most often is time.

Technical Problems Associated with Stress Multiple Gated Acquisition

Patient movement is one of the biggest problems associated with conducting this test. It is preferable to have a stress table or bicycle that can be bolted to the floor. The stress equipment must be designed to minimize chest movement. Velcro straps for the chest, shoulder pads, and hand grips to help stabilize the patient are some examples.

Patient cooperation is essential. The patient must understand what is expected of them and be fully able to cooperate. Changes in heart rate during acquisition are unavoidable during exercise. There will be a drop off in counts in the later frames of each acquisition.

An Alternate Acquisition Method

In some clinics, only one acquisition at peak exercise is acquired, instead of a series of exercise acquisitions at increasing workloads. First, a resting study is acquired. After the resting acquisition, the patient is asked to pedal at a constant number of revolutions per minute. The workload on the pedals is increased until the patient's heart rate reaches 85% of its maximum predicted value. At that point, an exercise acquisition is performed. When it is completed, a recovery study is acquired. This method does not take as long as the routine procedure. Because there are only three acquisitions, the patient may be able to tolerate the study better and the time per acquisition can be increased to 3–4 minutes.

Interpretation of Results

An EF is calculated for each acquisition. All of the walls of the heart should contract, and the septum should thicken. The wall motion of each acquisition is reviewed, particularly the rest, peak exercise, and recovery studies. In the normal study, the EF should increase and the wall motion should increase with exercise. A greater than 5 EF unit increase or a relative 10% increase in left ventricular EF from the baseline study is considered normal. Several factors may limit the increase in EF (Table 19.1). A fall in EF, no change, or less than a 5-unit increase is considered ab-

TABLE 19.1

Factors that Can Limit Ejection Fraction Increase with Exercise

Age older than 60 years
Valvular heart disease
Septal defects/shunts
Hypertension
Small ventricular volume/petite women

normal. The magnitude of the fall in EF increases as the severity of coronary artery disease worsens. Stress-induced left ventricular dysfunction represents the severity of the transient ischemia. During the recovery phase, the EF usually returns to baseline.

The sensitivity of stress exercise MUGA is between 85% and 90%. The sensitivity is highest for triple-vessel disease. However, the specificity is low, because exercise-induced left ventricular dysfunction is a common finding in many noncardiac diseases. Close concordance has been noted between exercise-induced regional wall motion abnormalities and reversible perfusion defects.

References

1. Gibbons RJ, Fyke FE III, Clements IP, Lapeyre AC III, Zinsmeister AR, Brown ML. Noninvasive identification of severe coronary artery disease using exercise radionuclide angiography. *J Am Coll Cardiol.* 1988;11:28–34.
2. Kelion AD, Banning AP, Ormerod OJ. Does exercise radionuclide angiography still have a role in clinical cardiac assessment? *J Nucl Cardiol.* 1999:6:540–546.
3. Wittry MD, Juni JE, Royal HD, Heller GV, Port SC. Procedure guidelines for equilibrium radionuclide ventriculography. *J Nucl Med.* 1997;38:1658–1661.
4. Iqbal A, Gibbons RJ, Zinsmeister AR, Mock MB, Ballard DJ. Prognostic value of exercise radionuclide angiography in a population-based cohort of patients with known or suspected coronary artery disease. *Am J Cardiol.* 1994;74:119–124.

Chapter 20

First-Pass Study

A first-pass study can be used to determine the right and/or left ventriular ejection fraction (EF) at rest or under pharmacologic stress. Both right and left ventricular EF can be obtained from a single injection and data acquisition. The patient can be sitting upright or supine for the study. Patient cooperation is less problematic with the first-pass study because of the short imaging time. The study can be performed with a variety of radiopharmaceuticals (*1*). Serial studies can be performed with appropriate radiopharmaceuticals.

This chapter will cover the technical aspects of performing a first-pass study, method variations, pitfalls, and problems. To obtain accurate results, the procedure for acquisition and processing of first-pass data requires considerable attention to details. Assessment of cardiac shunts also will be discussed in this chapter.

Choice of Radiopharmaceutical

When performing only first-pass imaging, ^{99m}Tc-diethylenetriamine pentaacetic acid (DTPA) is a good radiopharmaceutical choice. ^{99m}Tc-DTPA is rapidly cleared from the circulation by the kidneys, making it a suitable agent for serial studies. ^{99m}Tc-TcO_4 also can be used when a first-pass study is acquired just before a multiple gated acquisition (MUGA) study performed by in vivo labeling of red blood cells (RBCs). A first-pass study also can be performed using RBCs tagged with ^{99m}Tc by the in vitro method using the UltraTag kit (Mallinckrodt, Inc., St. Louis, MO) if the volume to be injected is kept small (i.e., 1 mL or less) (*2*). A first-pass study can be acquired before myocardial perfusion imaging using ^{99m}Tc-sestamibi or ^{99m}Tc-tetrofosmin. A dose of 20–25 mCi is recommended for first-pass imaging, regardless of the radiopharmaceutical used. The high dose is needed to obtain adequate statistics for a diagnostic study. For this reason, ^{201}Tl is not a suitable radiopharmaceutical for first-pass imaging (*3*).

Resting First-Pass Study Procedure

Choice of Collimator

The imaging time per frame is very short with a first-pass study (<50 msec). With a standard gamma camera, a high-sensitivity collimator is absolutely essential to obtain adequate statistics. Some equipment manufacturers have ultra-high-sensitivity collimators for first-pass work.

Patient Positioning and Preparation

Extensive patient preparation is not routinely required for this test. The patient usually is positioned supine on the imaging table. Although it is not a requirement for first-pass studies, R wave data can be collected with the acquisition. R waves can be useful to help identify the end diastolic (ED) frames in the study. The placement of the electrocardiographic (ECG) leads is the same as described in chapter 17.

Camera Positioning

The camera should be positioned as close to the chest as possible to maximize count rate detection and optimize resolution. For this reason, the anterior view is often used, particularly when both right and left ventricular EF are to be calculated from a single acquisition. When only one side of the heart is being evaluated, oblique positioning is also appropriate and gives a better delineation of the selected ventricle. A right anterior oblique view would be used for right ventricular EF. A left anterior oblique view would be used for the left ventricular EF.

With all positioning, the heart must be centered in the field of view. The great vessels must be included in the field of view as well. Figure 20.1 is a diagram of the anatomical landmarks used to ensure proper positioning for the first-pass study. When using a large-field-of-view camera, some zoom is needed to reduce the field of view to include only the area of the chest described in Figure 20.1.

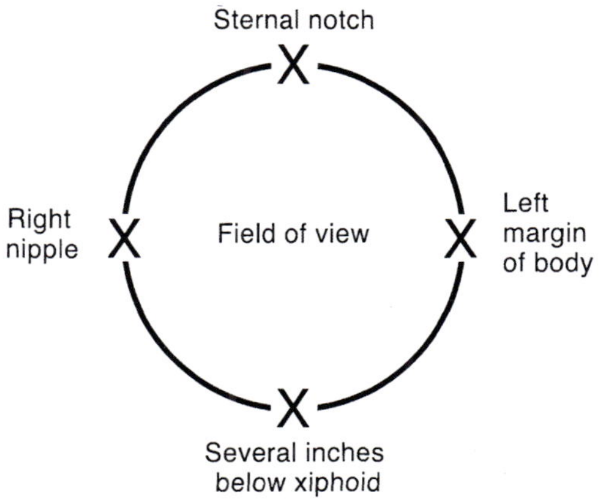

Figure 20.1 Anatomical landmarks used to position a first-pass study.

Injection Technique

The injection technique is a critical part of the procedure. To obtain a sharp delineation of each chamber of the heart, a small bolus of activity must be injected. The dose for first-pass imaging must be contained in a volume of 1.0 mL or less. For the left ventricular EF, a bolus of 0.4 mL or less is preferable.

The injection can be given via the external jugular vein or through an antecubital vein. When the injection is to be given through an antecubital vein, the median basilic vein in the right arm provides the most direct route to the superior vena cava (SVC) (*1*).

For both sites, an intravenous catheter with a short line should be established, using a 19–20-gauge needle. The line should be connected to a three-way stopcock. The radiopharmaceutical, followed by 3–5 mL saline, is injected rapidly using the three-way stopcock.

The radiopharmaceutical can also be administered by direct injection into the antecubital view using a Velcro blood pressure cuff as a tourniquet. The cuff should be inflated above the systolic pressure. After injecting the radiopharmaceutical, the cuff is quickly removed and the arm elevated (*4*).

Acquisition of Data

A 20% energy window is used. A 64 × 64 matrix is adequate for both resolution and image statistics. The acquisition is started immediately when the saline flush is pushed or when the blood pressure cuff is removed. The image acquisition time per frame can be individually determined based on the patient's heart rate. Images are acquired routinely at 25 msec/frame for 30 seconds. In this case, the total acquisition is 1,200 frames (*3,5*).

Data Processing

Data processing includes bolus assessment and right and/or left ventricular EF calculation. Both bolus assessment and EF calculation require drawing precise regions of interest (ROIs). ROI selection is most often performed manually, although automated software for first-pass ROI selection is available. The acquisition data are reviewed for the time interval during which the bolus is passing through the structures of interest, including the SVC, right ventricle, and left ventricle.

Bolus Assessment

The time frame during which the bolus travels through the SVC should be reviewed first. All frames showing the SVC are added together. Using this composite image, an ROI is drawn around the SVC. A time–activity curve for the SVC is generated. If the bolus is fragmented or if the transit time is prolonged, the study should be repeated. For a right ventricular EF, the full width at half maximum (FWHM) of the SVC curve should be 2–3 seconds. For a left ventricular EF, it should be 1 second or less (*3*). When right and left ventricular EFs are being assessed, a FWHM of 1.0–1.5 is adequate.

Right and/or Left Ventricular Ejection Fraction Calculation

The images are reviewed to determine the time frames during which the bolus is traveling through each side of the heart. The bolus traverses through the right heart to the pulmonary circulation and then to the left heart. Therefore, the time frame during which the bolus is seen in the right ventricle will be shorter than the time frame in which the bolus is seen in the left ventricle.

To flag the right ventricle, a composite image is made of the bolus passing through it. This composite image includes all the frames, starting with the frame in which activity is first seen in the right side of the heart through to the frame in which the outflow to the pulmonary circulation is first seen. When drawing the ROI over the right ventricle, careful attention should be given to the separation between the right atrium and right ventricle as well as to the outflow to the pulmonary circulation.

Frames in which the left heart are best seen are used to flag the left ventricle. When flagging the left ventricle ROI, care should be taken not to include any portion of the left atrium, right ventricle, or great vessels.

A time–activity curve of the bolus in the right or left ventricle is generated using the original acquisition data (25 msec/frame). For a right ventricular EF, a time–activity curve is generated using the right ventricle ROI and the time frame during which the bolus is traveling through the right side of the heart. The time frame for a right ventricle curve begins when the activity is first seen in the SVC and ends when the activity is seen leaving the right ventricle to the pulmonary circulation. For the left ventricular EF, a time–activity curve is generated using the left ventricle ROI and the time frame during which the bolus is traveling through the left ventricle. These time–activity curves will have peaks and valleys corresponding to ED and ES. The EF is calculated from one or more individual beats (Fig. 20.2).

The EF for a ventricle can be calculated from a single ROI for that ventricle. For more precise EF calculations, it is possible to flag ED, ES, and background ROIs (Fig. 20.3). Software is available to assist with the selection of the ED and ES frames for flagging. Background ROIs should not touch or include the heart or great vessels.

A variety of other software options and tools, such as phase analysis and amplitude, are available but vary from vendor to vendor. The transit time is the time it takes for the bolus to travel from the right to the left side of the heart. The pulmonary transit time is the time from the 75% point of the right ventricular peak count rate to the left ventricle peak. The normal pulmonary transit time is considered to be 5.7 seconds (*6*).

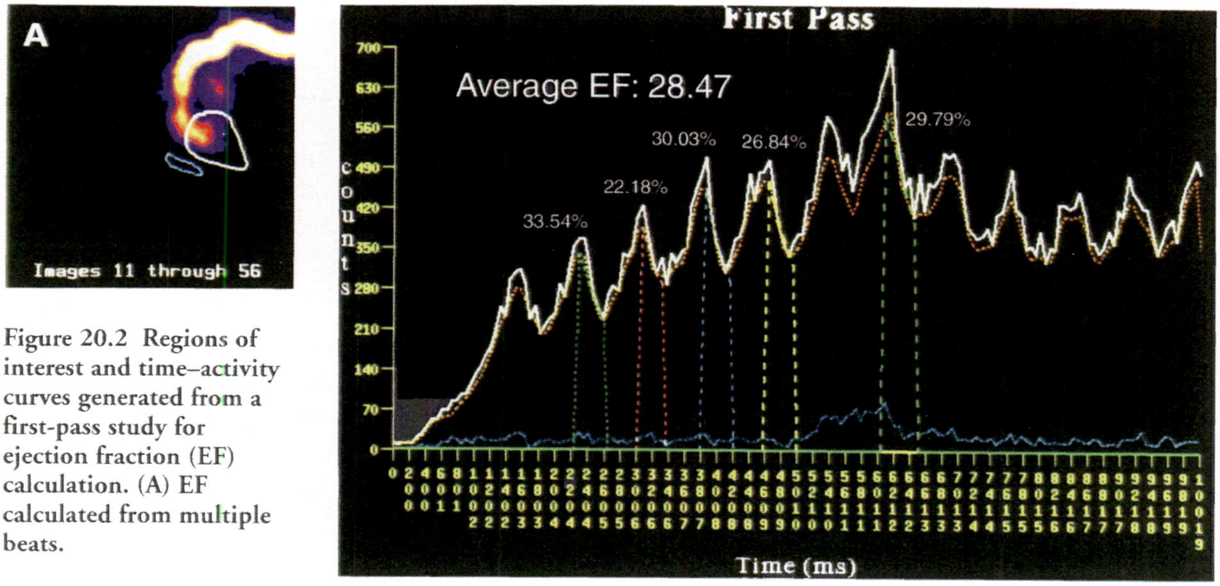

Figure 20.2 Regions of interest and time–activity curves generated from a first-pass study for ejection fraction (EF) calculation. (A) EF calculated from multiple beats.

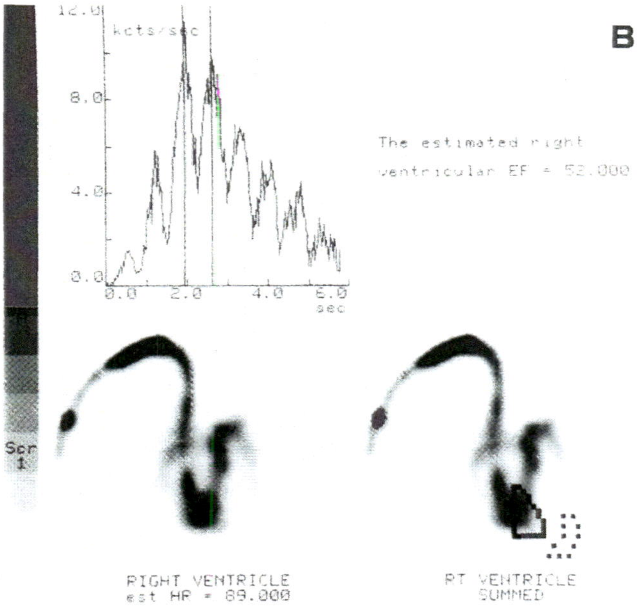

(B) EF calculated from a single beat.

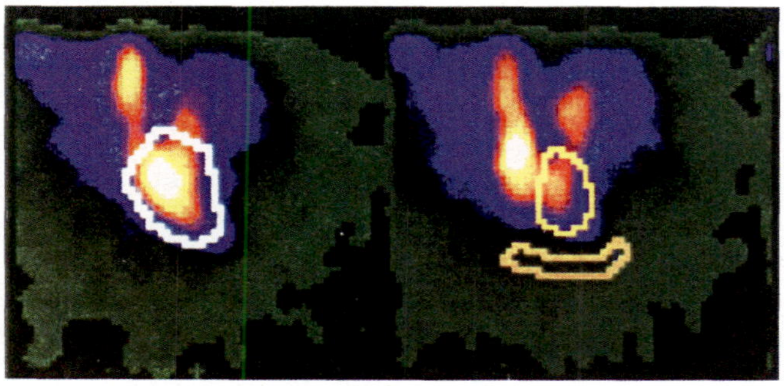

Figure 20.3 End diastole, end systole, and background regions of interest.

First-Pass Imaging at Peak Stress

A first-pass study can also be acquired at peak stress. The patient can be positioned upright or supine. The injection technique and imaging procedure are much the same as described previously. The stress method can be exercise or pharmacologic. Small-field-of-view multicrystal detector systems designed specifically for first-pass work are not in common use today. These systems worked well with treadmill exercise. As described in chapter 19, the major problem with bicycle and stress table exercise is patient movement.

Dobutamine is an alternative to exercise (*7*). The protocol for dobutamine stress would be similar to that described in chapter 4.

Cardiac Shunt Evaluation

Cardiac shunt refers to the movement of blood across the septum. A shunt may occur in the atrial or ventricular septum. The movement of blood may be right to left or left to right. Shunts are most often congenital defects that will eventually close on their own as patients mature. If a shunt does not close spontaneously, surgery may be indicated. Quantification of cardiac shunts is best accomplished by cardiac catheterization, which is an invasive procedure. Shunt evaluation is often performed on pediatric patients, and radiopharmaceutical doses must be adjusted accordingly. Nuclear medicine procedures can be useful in preliminary shunt evaluation. Ultrasound and doppler imaging are currently most often used for shunt evaluation.

Right-to-Left Shunts

A right-to-left shunt will result in cyanotic heart disease. A detection of a right-to-left shunt is made by review of cardiac first-pass study images. If a right-to-left shunt

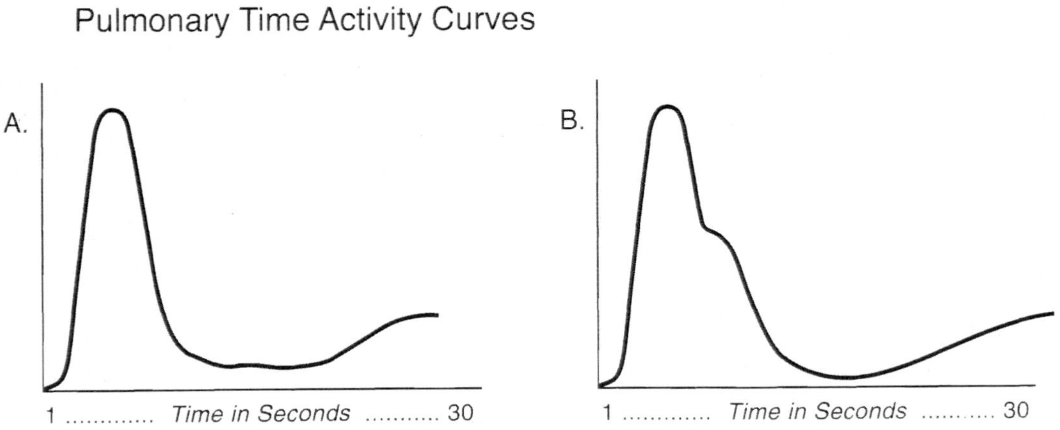

Figure 20.4 Pulmonary time–activity curves. (A) Normal curve. (B) Curve demonstrating a left-to-right shunt.

exists, there will be a premature visualization of the aorta. The aorta will visualize before the lungs. Also, the right heart may seem enlarged in comparison with the left. A right-to-left cardiac shunt can be quantitated using ^{99m}Tc macro-aggregated albumin (MAA) (*8,9*). This method requires anterior and posterior total body imaging after the administration of the ^{99m}Tc-MAA. Calculation of the percentage of right-to-left shunting is made by comparing the total body counts and total lung counts.

Left-to-Right Shunts

If a left-to-right heart shunt exists, blood will return prematurely to the pulmonary circulation. Detection of the presence of a left-to-right shunt is made by evaluation of a pulmonary time–activity curve generated from a first-pass study. The entire chest, including lungs and heart, should be included in the field of view. A good bolus injection as previously described is required. Images are acquired 0.5 seconds apart for 30 seconds.

An ROI is drawn around one or both lungs. The lung ROI should not include any heart structures or great vessels. A second ROI is drawn around the SVC. Two time–activity curves are generated, one for each ROI.

The SVC curve is used to assess the quality of the bolus. If the bolus was not good, a left-to-right shunt may not be apparent when reviewing the pulmonary time–activity curve. The peak of the SVC curve will have an FWHM of 3 seconds or less with a good bolus injection.

A normal pulmonary time–activity curve will have a single large peak, followed by a small peak some time later. The small peak corresponds to the second pass of activity through the lungs and comes from the return of activity from the systemic circulation.

If a left-to-right shunt exists, the initial peak of the pulmonary time–activity curve will have a bump on its downslope. The bump corresponds to activity that bypassed the systemic circulation and prematurely returned to the lungs (Fig. 20.4). Several methods have been described to quantitate left-to-right shunts and include the C2/C1 method, the gamma variate method, and area ratio method (*8,10,11*).

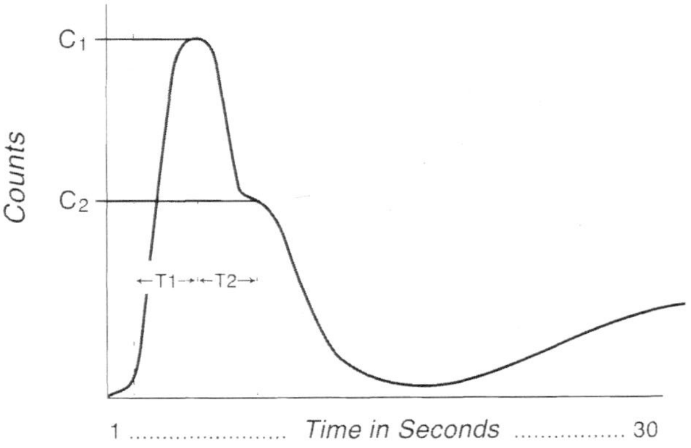

Figure 20.5 C2/C1 method for quantification of left-to-right shunts. T1 = time interval from the beginning of the rise of the pulmonary curve to the maximum count; T2 = time interval equal to T1 following T1; C1 = count level at the end of T1; and C2 = count level at the end of T2.

In the C2/C1 method, the counts at the end of two distinct time intervals are compared. The first time interval (T1) is the time from the beginning of the rise of the curve to the peak count rate (C1). T2 is the time interval equal to T1 and immediately after T1. C2 is the count level at the end of T2. Figure 20.5 describes the C2/C1 method. The C2/C1 ratio will be elevated when a right-to-left shunt exists. Each clinic can establish its own normal values.

References

1. Berger HJ, Matthay RA, Pytlik LM, Gottschalk A, Zaret BL. First-pass radionuclide assessment of right and left ventricle performance in patients with cardiac and pulmonary disease. *Semin Nucl Med.* 1979;9:275–295.
2. Stallings LE, Cheng KT, Spicer KM, Wiegner R, Humphries JK. Ultratag RBC kit for combined cardiac first-pass and multigated acquisition studies. *J Nucl Med Technol.* 1997;25:44–48.
3. Garcia EV. Imaging guidelines for nuclear cardiology procedures. Part I. *J Nucl Cardiol.* 1996;3:G3–G46.
4. Klingensmith WC, Eshima D, Goddard J, eds. *Nuclear Medicine Procedure Manual.* Englewood, CO: Wick Publishing Co.; 2000:6-1–6-3.
5. Ellis JH, Steele PP. Value of combined hemodynamic and radiocardiographic studies in acute respiratory failure. In: Serafini AN, Gilson AJ, Smoak WM, eds. *Nuclear Cardiology.* New York, NY: Plenum Publishing Corp.; 1977:187–201
6. Nichols K, DePuey EG, Gooneratne N, Salensky H, Friedman M, Cochoff S. First-pass ventricular ejection fraction using a single-crystal nuclear camera. *J Nucl Med.* 1994;35:1292–1300.
7. Iftikhar I, Koutelou M, Mahmarian JJ, Verani MS. Simultaneous perfusion tomography and radionuclide angiography during dobutamine stress. *J Nucl Med.* 1996;37:1306–1310.
8. Gelfand MJ, Hannon DW. Pediatric nuclear cardiology. In: Gerson MC, ed. *Cardiac Nuclear Medicine.* 2nd ed. New York, NY: McGraw-Hill, Inc.; 1991:551–590.
9. Gates GF. Right-to-left shunts. In: Gelfand MJ, Thomas SR. *Effective Use of Computers in Nuclear Medicine.* New York, NY: McGraw-Hill Inc.; 1988:288–300.
10. Gelfand MJ, Hannon DW. Left-to-right shunts. In: Gelfand MJ, Thomas SR. *Effective Use of Computers in Nuclear Medicine.* New York, NY: McGraw-Hill Inc.; 1988:271–287.
11. Madsen MT, Argenyi E, Preslar J, Grover-McKay M, Kirchner PT. An improved method for the quantification of left-to-right cardiac shunts. *J Nucl Med.* 1991;32:1808–1812.

Chapter 21

Cardiac PET Imaging

Positron emission tomography (PET) is a growing aspect of nuclear medicine imaging and utilizes positron-emitting radionuclides. The most common applications of PET are in the fields of oncology, cardiology, and neurology. As a result of widely accepted clinical applications, PET radiopharmaceuticals have become widely available. Myocardial PET imaging is used primarily for evaluation of perfusion and assessment of myocardial viability. A review of the clinical applications of PET in cardiology is presented here and followed by the major radiopharmaceuticals and procedures for assessing myocardial perfusion and viability.

Clinical Applications of PET in Cardiology

PET offers several advantages for cardiac imaging. PET imaging provides higher resolution than SPECT. PET studies acquire a much higher count density than does SPECT, resulting in better quality images. Attenuation correction of cardiac PET is routinely performed. Pharmacologic stress and rest perfusion imaging procedures can be completed in a shorter time than with SPECT stress and rest study. PET perfusion imaging is always performed with pharmacologic stress. Qualitative and quantitative assessment of myocardial perfusion and glucose utilization can be performed using PET.

The most common clinical application of PET in cardiology is the assessment of myocardial viability before revascularization procedures to predict outcomes in terms of functional recovery of the revascularized myocardium. Other clinical applications include early detection of coronary artery disease and detection of restenosis after revascularization procedures.

Clinical Implications

In the case of hibernating myocardium, left ventricular dysfunction may be inappropriately ascribed to necrosis and scar when, in fact, it may be completely reversible with revascularization (*1*). Poor left ventricular function was considered to be a relative contraindication for coronary artery bypass grafting. Myocardial viability assessment with PET is a better indicator of wall motion recovery after revascularization.

Myocardial Viability

Current concepts of myocardial viability assessment have evolved within the last 2 decades. Cardiac imaging modalities in nuclear medicine have played a leading role in assessing myocardial viability. The concepts of myocardial hibernation and stunning have become better understood. A reversible impairment of cardiac contractile dysfunction may result from myocardial stunning, silent ischemia, and/or hibernation.

Rahimtoola in 1989 defined myocardial hibernation as a state in which "contractile function declines chronically in response to a chronic reduction in blood flow, but improves or recovers if coronary blood flow is restored therapeutically" (*2*). In essence, this is an adaptive response to low blood flow, whereby oxygen demands are down-regulated for an extended period of time without cell death.

Hibernating myocardium can be detected by one of the following methods:

- Ionotropic stimulation;
- Myocardial perfusion imaging with ^{201}Tl; or
- Myocardial metabolic imaging with fluorodeoxyglucose (FDG) PET.

Braunwald and Klonar in 1982 defined myocardial stunning as a state in which "ischemia has been relieved and blood flow has been restored, while contractile function remains impaired initially and recovers only gradually" (*3*). In myocardial stunning, there is a decreased contraction in the face of normal perfusion. In other words, stunning is a perfusion–contraction mismatch. The major difference between myocardial stunning and hibernation is that the blood flow is near normal in stunned myocardium but is reduced in hibernating myocardium.

In hibernation and stunning, myocardial function is depressed but myocytes are viable. Therefore, myocardial dysfunction is reversible and, after restoration of myocardial perfusion, improvement in contractile function occurs.

Radionuclides Used for Cardiac PET

Radionuclides used for PET imaging are positron emitters. The radioactive decay process involves the transformation of a proton into a neutron and positively charged electron called a positron (β^+). The neutron remains in the nucleus of the atom, and the positron is emitted with kinetic energy. The amount of kinetic energy carried by a positron is different for each positron-emitting radionuclide. The more kinetic energy a positron has, the farther

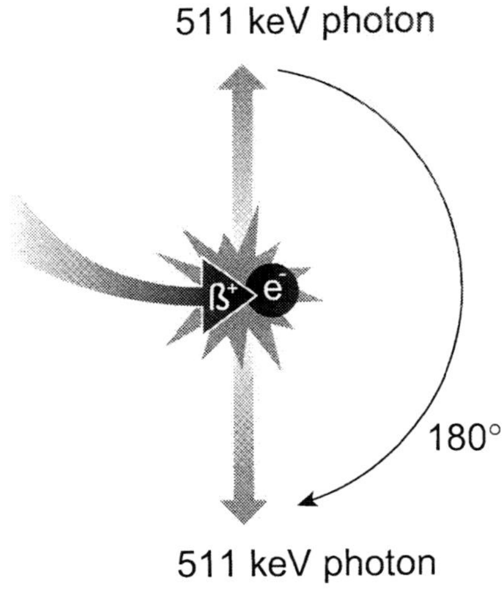

Figure 21.1 Diagram of positron annihilation.

away from its origin it will travel. When the positron loses all its kinetic energy and comes to rest, it will combine with an electron and they will annihilate. The mass of each particle (the electron and the positron) is equivalent to 511 keV of energy. Each annihilation reaction results in a pair of 511-keV photons emitted in opposite directions exactly 180° apart (Fig. 21.1).

In PET imaging, the two photons are detected at the same time by two detectors 180° apart. This is known as coincidence detection. The unique characteristic of coincidence imaging is the pair of 511-keV photons detected 180° apart.

For cardiac PET imaging, the most commonly used radionuclides are fluorine-18 (^{18}F), rubidium-82 (^{82}Rb), nitrogen-13 (^{13}N), and oxygen-15 (^{15}O). The half-life and kinetic energy of each of these radionuclides are listed in Table 21.1. Carbon-11 (^{11}C) is a radionuclide used less frequently for cardiac PET. All of the radionuclides listed, with the exception of ^{82}Rb, are cyclotron produced. ^{82}Rb is generator produced. When the radionuclide half-life is very short, the production source must be close to the PET scanner. The half-life can be a limiting factor in the use of some of these radionuclides. The more kinetic energy a positron carries, the farther away from its origin it will travel. The lower the kinetic energy, the better the resolution of the images.

TABLE 21.1

PET Radionuclides

Nuclide	Half-life (min)	Approx keV of the ß+ before annihilation*
^{18}F	110 minutes	249
^{82}Rb	75 seconds	1,520
^{13}N	10 minutes	492
^{15}O	122 seconds	725
^{11}C	20 minutes	386

* Each ß+ yields a 511 keV photon after annihilation.

TABLE 21.2

PET Radiopharmaceutical Doses

Agent	Dose
^{13}N-ammonia	18–20 mCi
^{15}O-water	≅10 mCi
^{82}Rb-chloride	50–60 mCi
^{18}F-fluorodeoxygloucose	10 mCi

Radiopharmaceuticals and Related Indications

The most commonly used radiopharmaceuticals for cardiac PET imaging are ^{18}F-FDG, ^{82}Rb-chloride, ^{13}N-ammonia, and ^{15}O-water. Myocardial perfusion can be assessed with ^{13}N-ammonia, ^{15}O-water, and ^{82}Rb-chloride (*4*). ^{82}Rb is also useful in assessing the integrity of the myocardial cell membrane. Myocardial metabolism (glucose uptake and phosphorylation) is assessed with ^{18}F-FDG. ^{11}C-acetate can be used for assessing oxidative metabolism. Typical doses of the common cardiac PET agent are listed in Table 21.2

Perfusion Agents

^{13}N-ammonia is cleared rapidly from the blood and diffuses into myocardial cells. Some of the tracer will be synthesized into ^{13}N-glutamine and become metabolically trapped. Myocardial uptake is proportional to the blood flow. ^{13}N-ammonia is a good myocardial perfusion agent for the detection of coronary artery disease. The half-life of ^{13}N-ammonia is 10 minutes. Use of this radiopharmaceutical requires a cyclotron in relatively close proximity to the PET scanner. The short half-life makes it possible to do multiple studies in a short time (*5–8*).

^{15}O-water is a freely diffusible tracer, and its uptake in the myocardium does not vary with flow rate or change in metabolic milieu of the myocardium. Because ^{15}O is a freely diffusible tracer, there will be a high concentration of ^{15}O in the blood. Correction must be made for the high level of blood activity. The target-to-background ratio with ^{15}O water is not as good as that with ^{13}N-ammonia or ^{82}Rb (*1,5–7,9–11*).

^{15}O water has a half-life of 2 minutes. It is possible to do multiple sequential studies with ^{15}O-water with minimal patient radiation exposure. ^{15}O-water use requires an onsite cyclotron. It is not routinely used for myocardial perfusion imaging.

^{82}Rb is a potassium analog and taken up by myocytes by a sodium potassium adenosine triphosphate (ATP) pump mechanism (*9*). Because of this energy-dependent active transport mechanism, ^{82}Rb can be used to evaluate myocardial cell membrane integrity and hence myocardial viability (*12*). ^{82}Rb-chloride is rapidly cleared from the blood in a manner similar to potassium.

The kinetic energy of ^{82}Rb is higher than that of other commonly used PET radionuclides. Therefore, the resolution of all other cardiac radiopharmaceuticals is relatively superior to that of ^{82}Rb. However, the diagnostic quality of ^{82}Rb is just as good as that of other PET tracers.

^{82}Rb has a half-life of 75 seconds. It is generator produced. It is a daughter of strontium-82 (^{82}Sr), which has a half-life of 25 days. Quality control on the ^{82}Rb generator involves checking the elution daily for minimum levels of ^{82}Sr and/or ^{85}Sr, according to the manufacturer's instructions. In the clinical setting, the generator's useful life is approximately 1 month.

The patient dose of ^{82}Rb is administered by an infusion pump device connected directly to the generator. The automation of the system minimizes radiation exposure to the technologist. Because it is produced by an onsite generator, the radiopharmaceutical is conveniently available 24 hours a day. The ultrashort half-life allows sequential rest and stress imaging without any delay. A high volume of studies is needed to justify the cost of the generator. The use of the ^{82}Rb generator is the most convenient way to perform perfusion/viability PET imaging, particularly in places that do not have direct access to a cyclotron.

Myocardial Metabolic Imaging Agents

Fatty acids are normally the predominant fuel source for myocardial metabolism. The ischemic myocardium preferentially uses glucose in place of fatty acids, because it is more suited for anaerobic glycolysis.

^{18}F-FDG is the agent of choice for metabolic imaging. In the fasting state, FDG is taken up by myocardial cells in proportion to plasma glucose level. Once inside the cell, it is phosphorylated to FDG-6 phosphate but not further metabolized. The FDG-6 phosphate remains trapped in the cell for a sufficient amount of time to allow metabolic PET imaging. The FDG uptake in normal myocardium is usually good and fairly homogeneous. Infarcted and nonviable myocardium does not accumulate any significant amount of FDG. Ischemic myocardial segments usually demonstrate an increased FDG uptake. With a half-life of 110 minutes, ^{18}F-FDG is relatively easy to obtain and transport. It does not require an onsite cyclotron. The low kinetic energy of the ^{18}F positron yields high-resolution images of the heart (*13–20*).

Other tracers, such as ^{11}C-acetate, have also been used for evaluation of regional myocardial oxygen consumption, providing an insight to metabolic derangements (*21*).

Basic Principles of PET Imaging

A thorough review of the principles of PET imaging is beyond the scope of this text. The PET image acquisition and data reconstruction processes are significantly different from SPECT imaging (*22–24*). The following brief review of the basic principles is given to help the reader appreciate how PET imaging differs from conventional imaging with a gamma camera.

PET Cameras

A typical PET camera consists of multiple detectors arranged in a ring configuration. Crystals commonly used in PET cameras are bismuth germanate oxide (BGO), lutetium orthosilicate (LSO), gadolinium orthosilicate (GSO), and sodium iodide (NaI). In conventional PET camera design, the basic detection unit is arranged in blocks. The blocks contain the crystals, scored or divided in some manner to create the planes. Behind the crystals are a number of photomultiplier tubes (PMTs). The number of crystals and PMTs varies with the manufacturer and model. A series of blocks is arranged in a ring. The number of rings in the scanner determines the total axial field of view (FOV), ranging from 10–16 cm, depending on the make and model.

For example, a PET camera might be constructed with small BSO crystals (6-mm square). Each block could have 64 (eight rows × eight columns) of crystals. Behind each block of crystals are four PMTs. The ring would consist of eight blocks with a total of 512 crystals. As another example, an NaI PET scanner may use six large crystals arranged in a hexagon.

The PET camera detector system is equipped with coincidence circuitry. The axial FOV of a PET system is 10–16 cm, and the spatial resolution is approximately 5–10 mm. The gantry houses the multiple crystal detector system, coincidence circuitry, electronics, and patient positioning devices (Fig. 21.2). Because of coincidence cir-

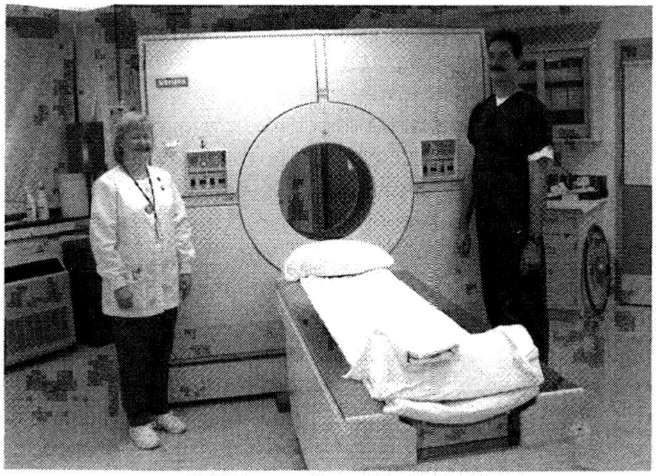

Figure 21.2 A PET scanner.

cuitry, collimators are not needed or used with PET cameras. PET cameras are designed with septa that prevent crosstalk across the rings. When the septa are in place in the FOV, crosstalk is limited. Imaging with septa is two-dimensional. Imaging without the septa increases the sensitivity in the FOV and allows for three-dimensional imaging. The PET camera comes with a bed, which is equipped with a device for automatic and/or manual positioning of the patient.

Data Acquisition and Processing

Before data acquisition, the patient must be positioned in the ring to ensure that all of the heart is in the FOV. This is done with a position scan. This step is especially crucial with small-FOV PET cameras. When the FOV is small and the heart is large, the entire heart may not fit in the FOV. In these situations, it is important that the myocardial segment in question be in the FOV. Portions of the heart not in question may be cut out of the FOV. The FOV of current generation PET cameras is large enough to accommodate even the largest heart.

The position scan can be acquired with a small amount of radiopharmaceutical, such as 20 mCi ^{82}Rb, or with a short attenuation scan. If the heart is outside the FOV on the position scan, the technologist must adjust the imaging table to get the heart in the FOV. A position scan is sometimes performed with a rectilinear scan made of three bed positions. The rectilinear scan is performed with the transmission source. Upon completion, the bed is moved to the appropriate position.

The PET scanner is equipped with a laser to check the patient's position before and after each acquisition. The patient's position should not change throughout the duration of the study. After positioning the heart in the FOV, the position of the laser light can be marked on the chest. These marks can be checked to ensure the patient did not move.

Once the patient is positioned, a static or dynamic scan of the target organ is acquired. The dynamic mode consists of multiple frames of data. The dynamic acquisition is used to assess changing tissue concentration of radiotracer.

Data acquisition involves recording coincidence events. A coincidence event is recorded each time two crystals 180° apart detect a 511-keV photon in the same narrow window of time. A line connecting the interactions in the two crystals is called a line of response. A line of response is recorded for each coincidence event. Numerous lines of response produce a raw data sinogram. The raw data are processed using iterative reconstruction into images of the distribution of activity in the organ of interest.

Four types of interactions or events can occur when the data are being acquired. These are true coincidence, random coincidence, single, and multiple events. A multiple event occurs when three events are detected at the same point in time. The correct line of response cannot be determined from a multiple event. Some of the elements that contribute to the degradation of PET images include scatter events, random events, camera dead time, and attenuation. For optimal imaging, the contribution of each should be minimized or corrected whenever possible.

Transmission Scan for Attenuation Correction

The PET camera is equipped with a transmission source, such as germanium-68 (^{68}Ge) or cesium-137 (^{137}Cs). The half-life of ^{68}Ge is 287 days. ^{68}Ge rod sources should be replaced approximately every 10 months. The transmission rod sources rotate around the patient, and a scan is acquired in a fashion similar to a computed tomography (CT) scan. Combined (or hybrid) CT/PET scanner systems do not need a transmission source; the attenuation correction is obtained from the CT transmission image.

The transmission scan measures the attenuation caused by various tissue densities and objects in the FOV. A patient is positioned in the PET camera FOV for a transmission scan before receiving the radiopharmaceutical. The transmission scan for attenuation correction can be acquired for either time or counts. When the transmission scan is acquired for counts, the time it takes to acquire the scan will vary, depending on the strength of the source. For example, an attenuation scan acquired with ^{68}Ge for 120 million counts can take as little as 10 minutes when the rod sources are new. When the sources need to be replaced, this scan can take as long as 30 minutes. The acquisition time of the transmission scan for attenuation correction acquired for counts is also dependent on the size of the patient. A blank transmission scan with nothing in the FOV is compared with the scan taken with the patient in the FOV to obtain the measured attenuation correction factors. Once the transmission data are acquired, the transmission source is deactivated and the patient is injected with a positron-emitting radiopharmaceutical for imaging and acquisition of emission data. The newer generation PET cameras are capable of acquiring transmission and emission data simultaneously.

PET Camera Quality Control

The basic performance measurements of a PET camera include spatial resolution, true sensitivity, count rate losses, randoms, uniformity, scatter, attenuation, and count rate linearity corrections.

Detector normalization and block set-up are performed on a monthly basis. Daily quality control requires acquiring a "blank" and checking for electronic drifts from original normalized values. Preventative maintenance is performed on a quarterly basis.

Imaging Protocols

Imaging protocols will vary slightly from institution to institution. The following protocols are given as examples.

Rest and Pharmacologic Stress Perfusion Imaging with ^{13}N-Ammonia

Patient Preparation

1. Instructions specific to pharmacologic stress method. Pharmacologic stress could be performed with dipyridamole, adenosine, or dobutamine.
2. Diabetics should take the usual dose of insulin or oral antidiabetic medication.
3. A pregnancy test must be obtained for women of childbearing age. If the pregnancy test is negative, the procedure can be performed.
4. Upon arrival for the appointment, a physician or technologist will explain the procedure to the patient and obtain a signed consent.
5. The patient should be asked to void, because the average imaging time for the study is 2 hours.
6. A 12-lead electrocardiograph (ECG) is connected for monitoring.
7. An intravenous line is inserted into the patient's arm for the administration of the pharmacologic stress agent.
8. An intravenous line is started on the opposite arm for the administration of the radiopharmaceutical and saline.

The patient's arms are above his or her head throughout the duration of the study.

Patient comfort is essential to minimize motion artifacts. The intravenous lines must be properly positioned for easy access and care during these procedures.

Acquisition Steps

1. Perform position scan.
2. Obtain transmission scan for attenuation correction.
3. Administer dose of ^{13}N-ammonia.
4. Acquisition of emission scan (rest scan) begins immediately after the administration of the radiopharmaceutical.
5. Patient remains in position for approximately 1 hour after injection to allow adequate decay before pharmacologic stress.
6. Upon completion of the decay time, the pharmacologic stress is performed.
7. Administer the second dose of ^{13}N-ammonia.
8. Acquire a second emission scan (stress scan) immediately after administration of radiopharmaceutical.

Rest and Pharmacologic Stress Perfusion Imaging with ^{82}Rb

Patient Preparation

The patient preparation would be similar to that described for the ^{13}N-ammonia procedure.

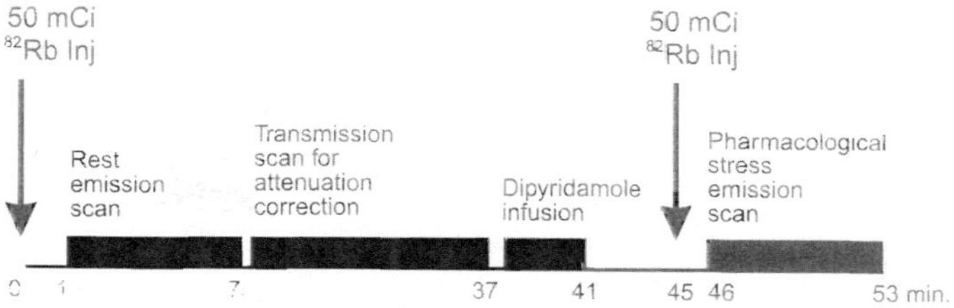

Figure 21.3 Rest–stress ^{82}Rb perfusion study protocol.

Acquisition Steps

1. Perform position scan.
2. Administer dose of ^{82}Rb via the generator infusion system.
3. Wait 70 seconds before starting acquisition after ^{82}Rb infusion is complete. (This step allows lung activity to clear before imaging.)
4. Obtain rest emission scan. (Rest emission scan is acquired for 6 minutes.)
5. Obtain transmission scan for attenuation correction after the completion of the rest emission scan.
6. Perform pharmacologic stress study.
7. Administer dose of ^{82}Rb via the infusion pump system.
8. Obtain stress emission scan 70 seconds after the ^{82}Rb infusion is completed. (Stress emission scan is acquired for 6 minutes.)

The total acquisition time for this study is approximately 1 hour. Figure 21.3 is a diagram showing the time and steps of this procedure. Approximately 20–40 million counts are acquired during each 6-minute acquisition.

Metabolic Imaging with ^{18}F-FDG

When evaluating myocardial viability with FDG, a resting perfusion study is performed first. The resting perfusion scan could be performed with ^{13}N-ammonia or ^{82}Rb as previously described. The resting perfusion study may have been performed with ^{201}Tl or a ^{99m}Tc-labeled myocardial perfusion agent. Any of these perfusion studies can be compared with metabolic images acquired with ^{18}F-FDG.

Patient Preparation

1. The patient eats a light, nonfat, high-carbohydrate breakfast or lunch.
2. Diabetics should take the usual dose of insulin or oral antidiabetic medication.
3. A pregnancy test must be obtained for women of childbearing years.
4. Upon arrival for the appointment, a physician or technologist will explain the procedure to the patient.
5. A 12-lead ECG is connected for monitoring.
6. A catheter is inserted into the patient's arm to monitor blood glucose and potassium levels as needed. An intravenous line is started on the opposite arm for the administration of the radiopharmaceuticals, saline and glucose, or insulin as needed.

Figure 21.4 Rest–stress ^{82}Rb perfusion scans. (A) Normal study.

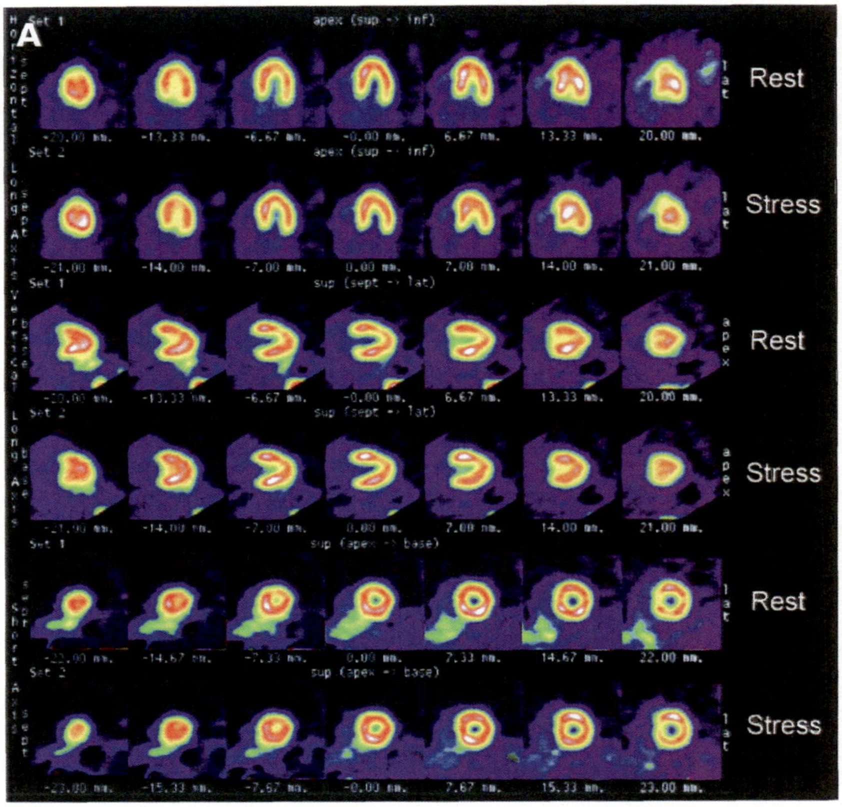

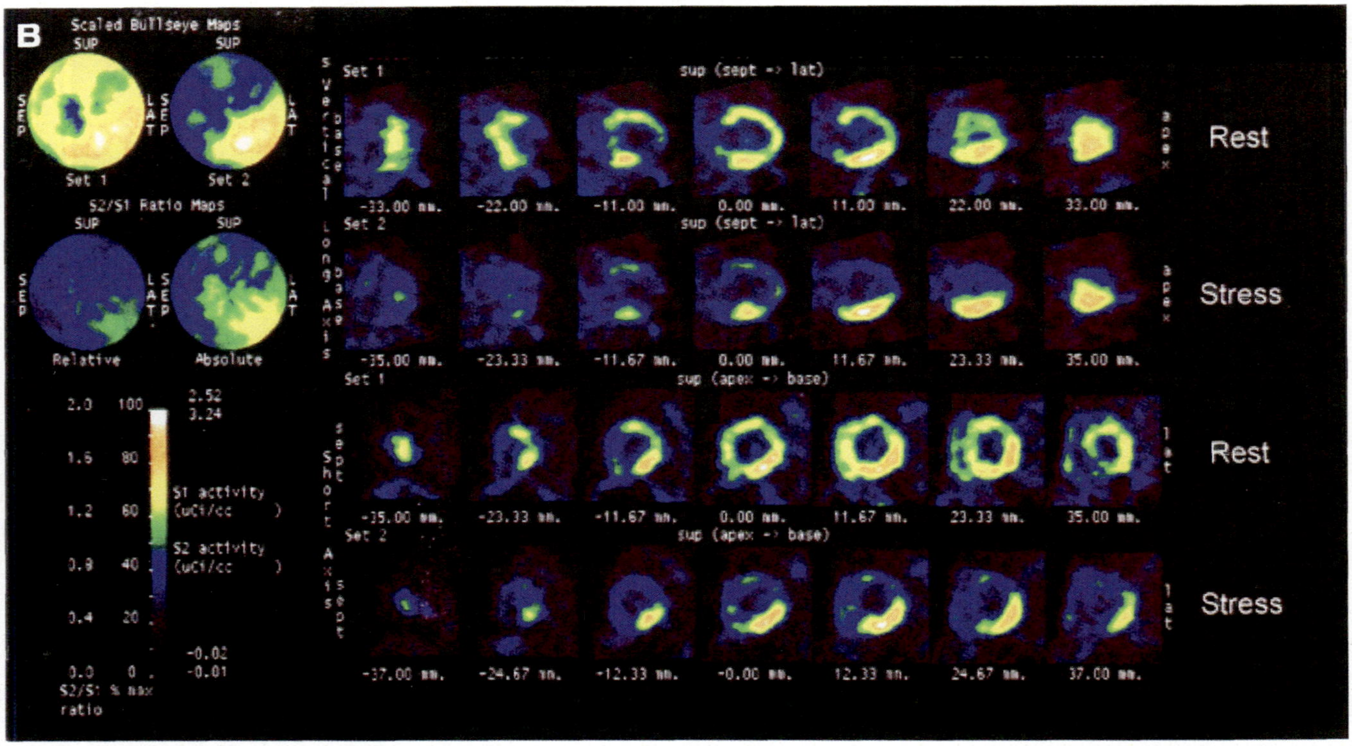

Figure 21.4 (B) Ischemia. This scan shows evidence of ischemia in the left anterior descending distribution involving the anterior, anterolateral, and anteroseptal regions of the left ventricle.

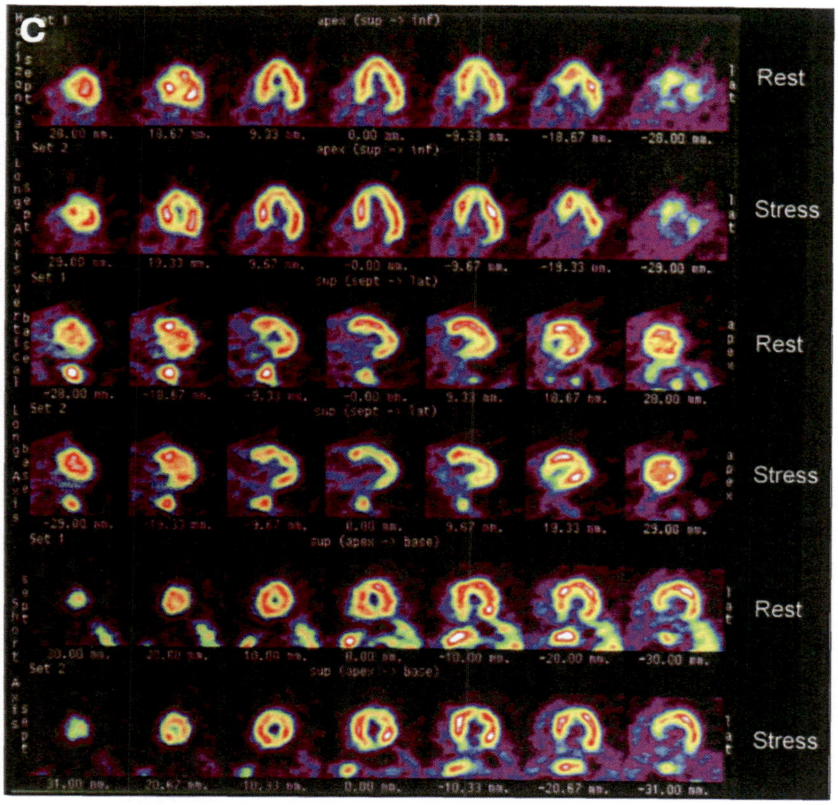

Figure 21.4 (C) Fixed defect. This scan shows a severe fixed defect in the inferior/inferobasal walls consistent with a scar from a previous myocardial infarction.

Monitoring and Adjustment of Blood Glucose Levels

An optimal blood glucose level is important for myocardial uptake of FDG in assessing myocardial viability. The blood glucose level should be between 120 and 160 mg/dL. If the blood glucose level is less than 120 mg/dL, an oral glucose loading with 100 gm glucose is usually sufficient.

Myocardial uptake of FDG, however, is unpredictable after oral glucose loading. Intravenous infusion of a glucose, insulin, and potassium (GIK) cocktail is found to be more predictable for glucose loading for higher myocardial FDG uptake. The composition and rate of infusion of the cocktail are:

Glucose 300 gm/L + regular human insulin 60 U/L + 40–60 mEq/L potassium infused at 1.5 mL/kg/hr.

In diabetic patients, the glucose levels and response to GIK infusion are often unpredictable. To achieve a satisfactory level of blood glucose, additional intravenous doses of regular insulin may be required. Usually 2.5 units of regular insulin are administered if blood glucose is between 160 and 180 mg/dL and 5 units if blood glucose levels are higher than 180 mg/dL.

Acquisition Steps

1. Perform position scan.
2. Obtain transmission scan for attenuation correction.
3. Administer ^{18}F-FDG.
4. Acquire emission data; multiple frame dynamic acquisition. Examples of frames acquired: 8 × 15, 2 × 30, 2 × 120, 1 × 180, 4 × 300, and 3 × 600 sec (total = 20 frames).
5. Obtain additional static images as needed, if the glucose uptake in the myocardium is not optimal.

Image Reconstruction

The acquisition data are attenuation corrected. The camera manufacturer will recommend an appropriate filter for a given study. If the patient is very obese, a smooth filter may be required to reduce the noise level (Fig. 21.5).

Short axis, horizontal long axis, and vertical long axis slices are constructed for each emission study. Polar tomograms and volume images are also constructed from the data.

Normal and Abnormal Scan Patterns: Image Review

Example rest–stress perfusion studies performed with ^{82}Rb and ^{13}N-ammonia are shown in Figures 21.4–21.6. All of these studies were performed with dipyridamole stress.

The normal physiologic uptake of FDG in the myocardium of a normal fasting adult is seen in the total body scan (Fig. 21.7). Example viability studies performed with ^{18}F-FDG are shown in Figure 21.8. For each sectional view, the resting perfusion scan is displayed first with the ^{18}F-FDG scan directly below. Figure 21.9 shows the effect of the blood glucose level on the viability study.

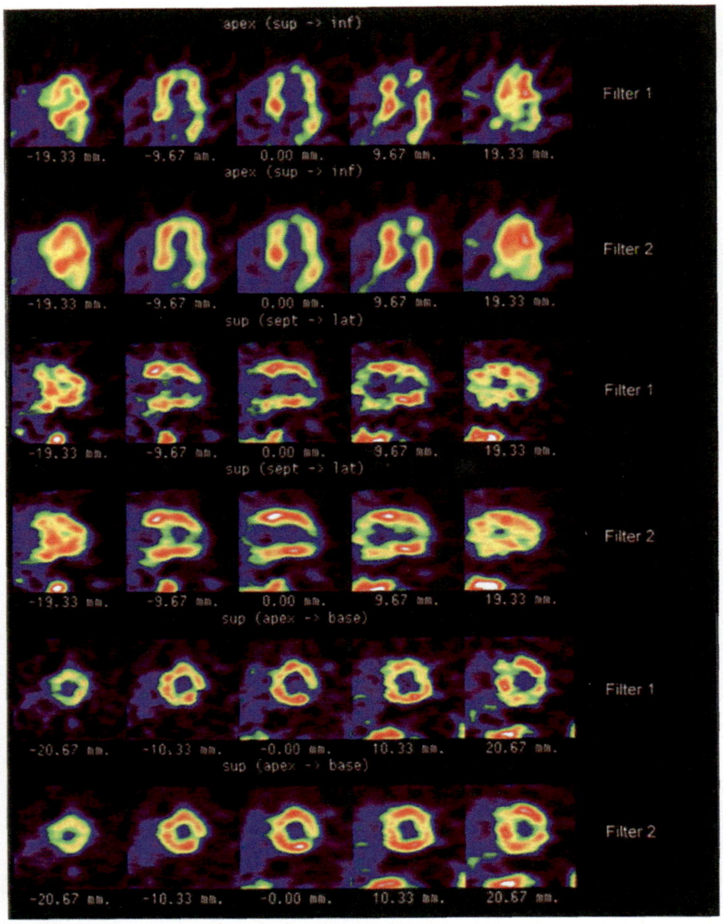

Figure 21.5 Rest ^{82}Rb study in an obese patient. Processed with two different filters.

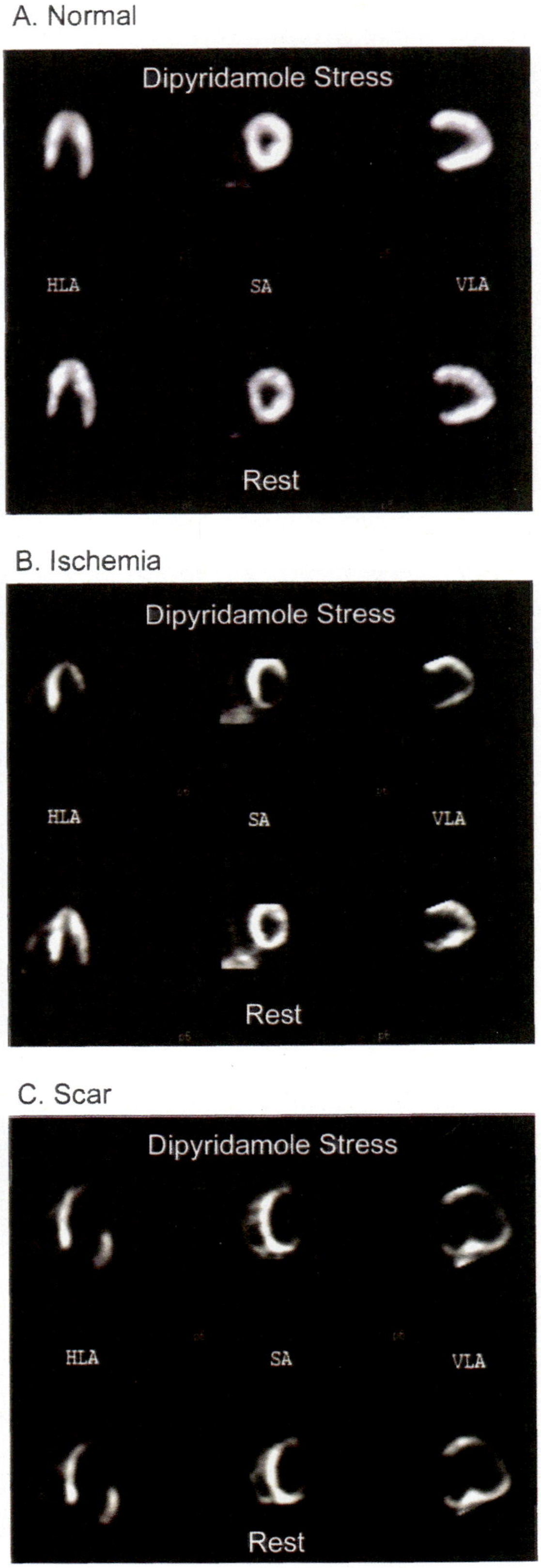

Figure 21.6 Rest–stress ^{13}N-ammonia perfusion scan. (A) Normal study. (B) These images show inferolateral/lateral wall ischemia. (C) Fixed defect. These images show a lateral wall scar.

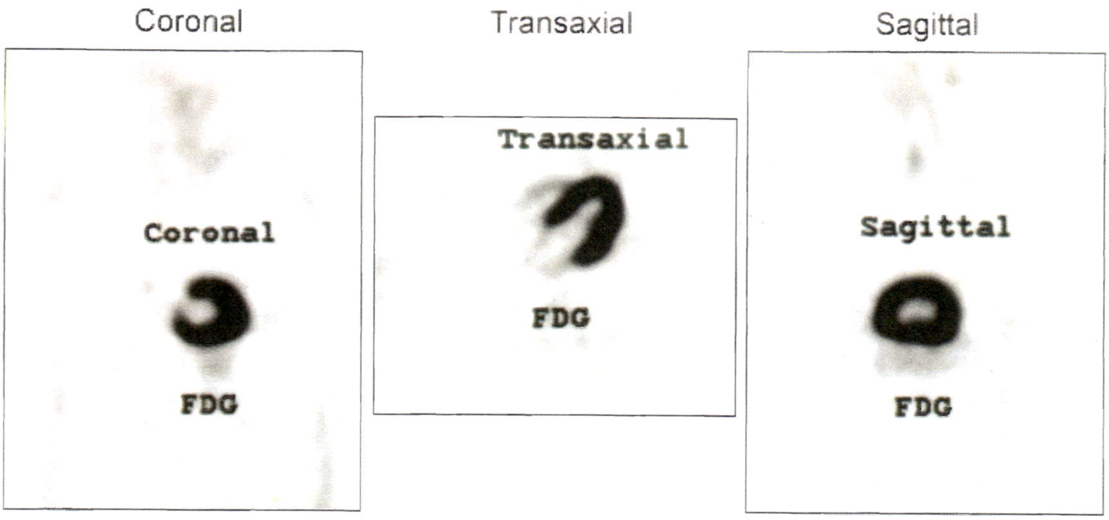

Figure 21.7 ^{18}F-fluorodeoxyglucose (FDG) whole-body scan. This whole-body scan shows the normal physiologic uptake of FDG in the myocardium of a healthy fasting adult.

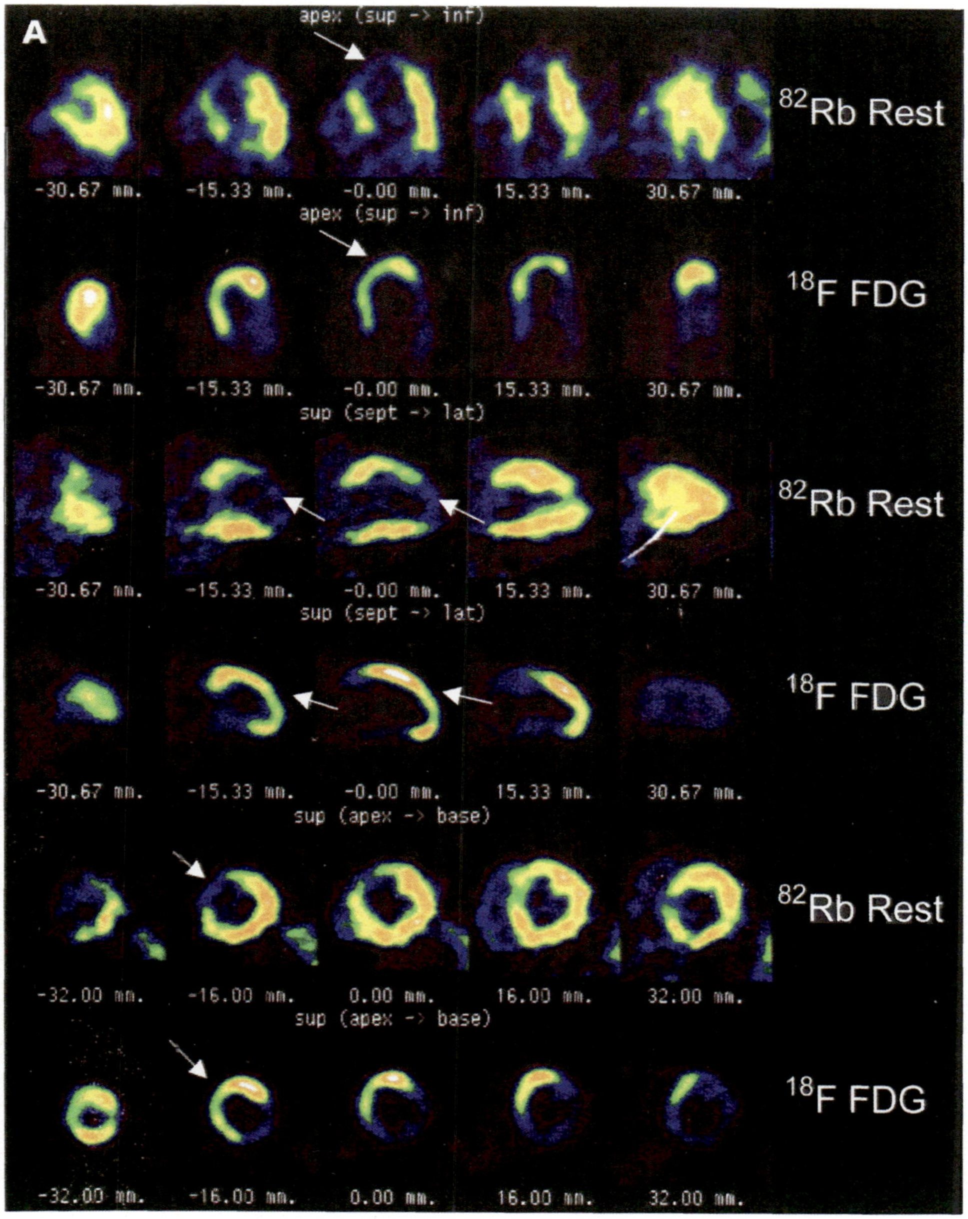

Figure 21.8 Viability studies. (A) Hibernating myocardium. The resting ^{82}Rb study (top line in each sectional view) shows an ischemic area involving the anterior, anteroapical, and anteroseptal regions of the left ventricle. These ischemic areas show good fluorodeoxyglucose uptake consistent with hibernating but viable myocardium.

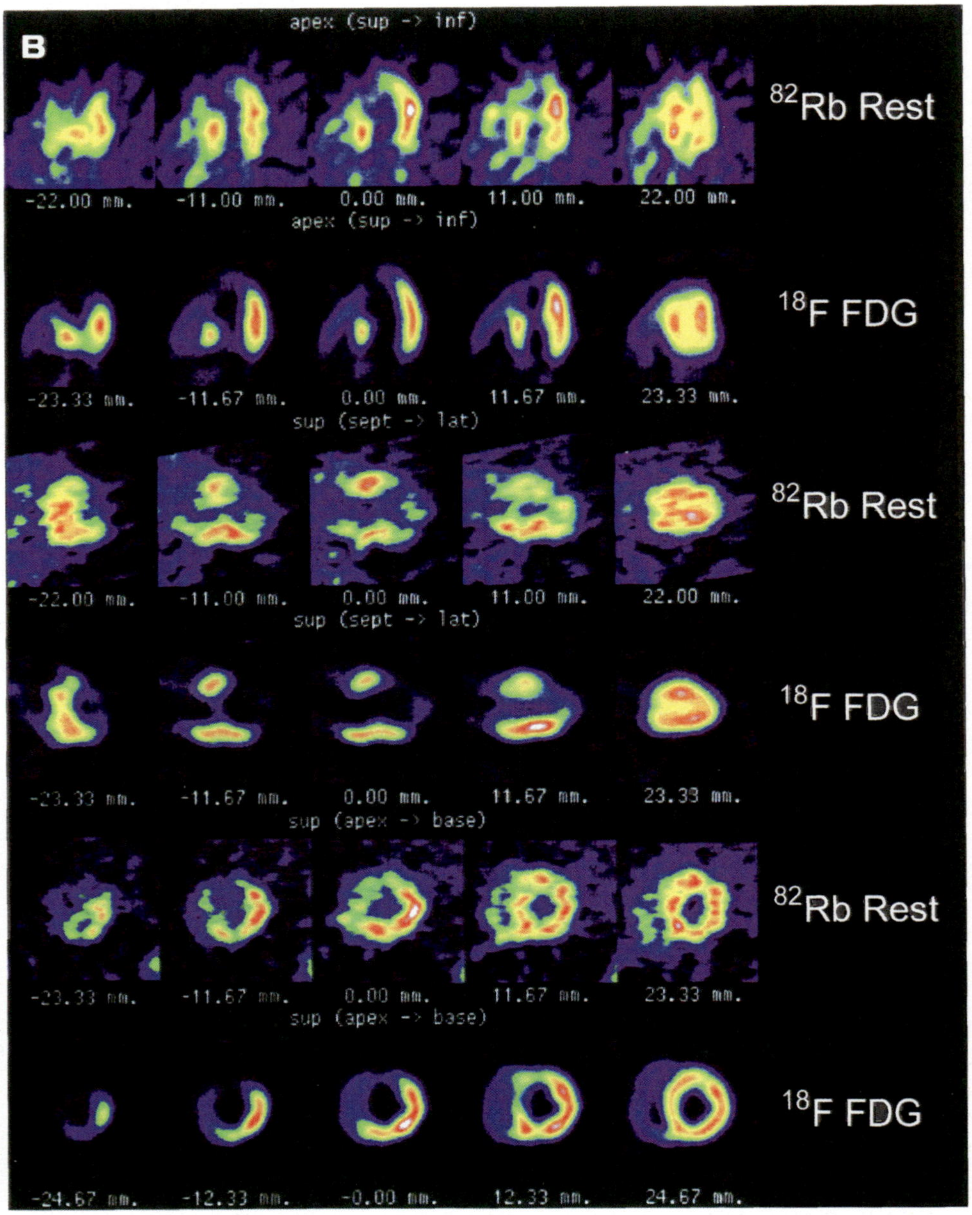

Figure 21.8 (B) Matching defect. Nonviability. This study shows a severe matching perfusion/metabolism defect involving the anteroseptal and apico-septal regions consistent with nonviability.

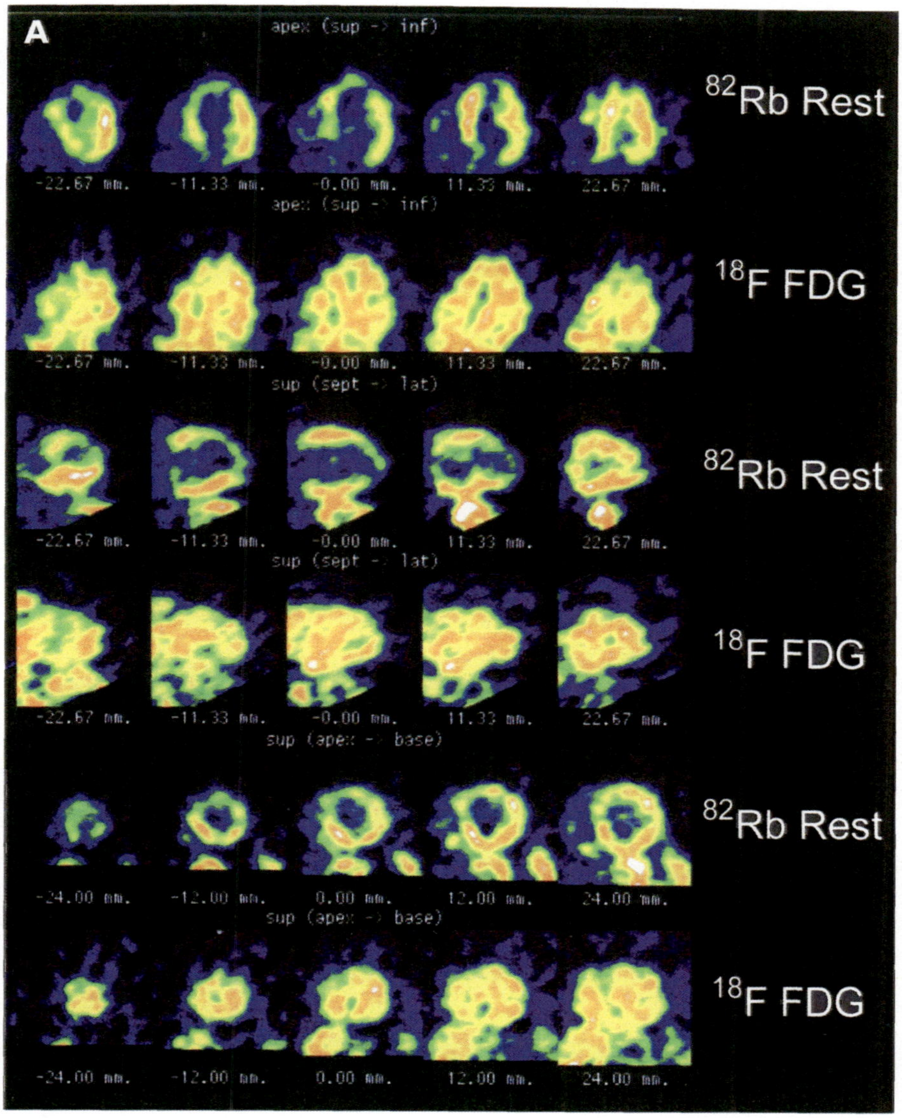

Figure 21.9 Viability study. Varying blood glucose levels. (A) The patient's blood glucose level was 239 mg/dL when this study was performed. There is poor inhomogeneous uptake of fluorodeoxyglucose in the myocardium.

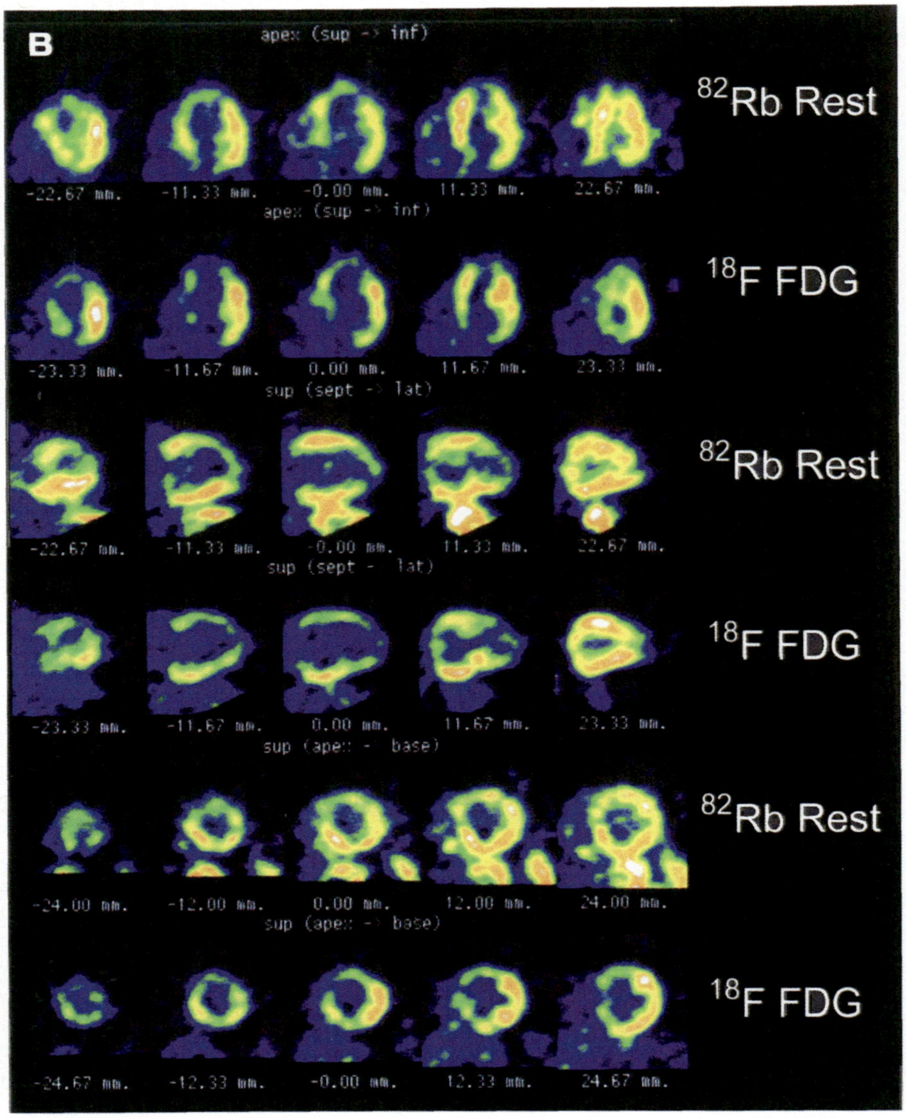

Figure 21.9 (B) These images of the same patient show improved FDG myocardial uptake after the blood glucose level was titrated down with insulin to 129 mg/dL.

References

1. Braunwald E, Rutherford JD. Reversible ischemic left ventricular dysfunction: evidence for the "hibernating myocardium". *J Am Coll Cardiol.* 1986;8:1467–1470.
2. Rahimtoola SH. The hibernating myocardium. *Am Heart J.* 1989;117:211–221.
3. Braunwald E, Kloner RA. The stunned myocardium: prolonged, postischemic ventricular dysfunction. *Circulation.* 1982;66:1146–1149.
4. Schwaiger M. Myocardial perfusion imaging with PET. *J Nucl Med.* 1994;35:693–698.
5. Schelbert HR, Phelps ME, Huang SC, et al. N-13 ammonia as an indicator of myocardial blood flow. *Circulation.* 1981;63:1259–1272.
6. Bol A, Melin JA, Vanoverschelde JL, et al. Direct comparison of ^{13}N ammonia and ^{15}O estimates of perfusion with quantification of regional myocardial blood flow by microspheres. *Circulation.* 1993:87:512–525.
7. Muzik O, Beanlands RS, Hutchins GD, Mangner TJ, Nguyen N, Schwaiger M. Validation of nitrogen-13-ammonia tracer kinetic model for quantification of myocardial blood flow using PET. *J Nucl Med.* 1993;34:83–91.
8. Nienaber CA, Ratib O, Gambhir SS, et al. A quantitative index of regional blood flow in canine myocardium derived noninvasively with N-13 ammonia and dynamic positron emission tomography. *J Am Coll Cardiol.* 1991;17:260–269.
9. Gould KL. PET perfusion imaging and nuclear cardiology. *J Nucl Med.* 1991;32:579–606.
10. Walsh MN, Bergmann SR, Steele RL, et al. Delineation of impaired regional myocardial perfusion by positron emission tomography with $H_2{}^{15}O$. *Circulation.* 1988;78:612–620.
11. Williams BR, Mullani NA, Jansen DE, Anderson BA. A retrospective study of the diagnostic accuracy of a community hospital-based PET center for the detection of coronary artery disease using rubidium-82. *J Nucl Med.* 1994;35:1586–1592.
12. Gould KL, Goldstein RA, Mullani NA, et al. Non-invasive assessment of coronary stenoses by myocardial perfusion imaging during pharmacologic coronary vasodilation. VIII. Clinical feasibility of positron cardiac imaging without a cyclotron using generator-produced rubidium-82. *J Am Coll Cardiol.* 1986;7:775–789.
13. Phelps ME, Hoffman EJ, Selin C, et al. Investigation of F-18 fluoro-2-deoxyglucose for the measurement of myocardial glucose metabolism. *J Nucl Med.* 1978;19:1311–1319.
14. Schwaiger M, Hicks R. The clinical role of metabolic imaging of the heart by positron emission tomography. *J Nucl Med.* 1991;32:565–578.
15. Ratib O, Phelps ME, Huang SC, Henze E, Selin CE, Schelbert HR. Positron tomography with deoxyglucose for estimating local myocardial glucose metabolism. *J Nucl Med.* 1982;23:577–586.
16. Brunken R, Schwaiger M, Grover-McKay M, Phelps ME, Tillisch J, Schelbert HR. Positron emission tomography detects tissue metabolic activity in myocardial segment with persistent thallium perfusion defects. *J Am Coll Cardiol.* 1987;10:557–567.
17. Schelbert HR. ^{18}F-deoxyglucose and the assessment of myocardial viability. *Semin Nucl Med.* 2002;32:60–69.
18. Dilsizian V, Bonow RO. Current diagnostic techniques of assessing myocardial viability in patients with hibernating and stunned myocardium. *Circulation.* 1993;87:1–20.
19. Mankoff D, Nader RG, Eisen HJ. Cardiac applications of positron emission tomography. *J Nucl Med Technol.* 1990:18:69–78.
20. Choi Y, Brunken RC, Hawkins RA, et al. Factors affecting myocardial 2 F-18 fluoro-2-deoxy-D-glucose uptake in positron emission tomography studies of normal humans. *Eur J Nucl Med.* 1993;20:308–318.
21. Armbrecht JJ, Buxton DB, Brunken RC, Phelps ME, Schelbert HR. Regional myocardial oxygen consumption determined noninvasively in humans with 1-^{11}C acetate and dynamic positron tomography. *Circulation.* 1989;80:863–872.
22. van der Wall EE, Blanksma PK, Niemeyer MG, et al., eds. *Cardiac Positron Emission Tomography.* Boston, MA: Kluwer Academic Publishers; 1995.
23. Schwaiger M, ed. *Cardiac Positron Emission Tomography.* Boston, MA: Kluwer Academic Publishers; 1996.
24. Turkington TG. Introduction to PET instrumentation. *J Nucl Med Technol.* 2001;29:4–11.

Chapter 22

Fundamentals of Electrocardiography

This section will cover the basics of electrocardiography (ECG) and will describe the basic principles of obtaining an ECG, the normal tracing, terminology, and common abnormalities that should be recognized. For a more extensive and comprehensive review of ECG, the reader should refer to the references listed and other suitable sources of information on this subject (*1–6*).

Conduction System Review

Impulse Formation

Each normal heart beat is a result of an electrical impulse that originates from the sinoatrial (SA) node. The SA node is a specialized area of nerve tissue located high in the right atrium (see chapter 3, Fig. 3.1). Normally, the SA node is the natural pacemaker controlling the heart rate. The SA node discharges electrical impulses 60–100 times per minute in a rhythmic fashion.

Impulse formation is an inherent property of cardiac muscle. Other areas of the heart have the ability to initiate impulses but do not do so normally or routinely. Electrical impulses that originate from any area of the heart other than the SA node cause ectopic beats. An ectopic site of impulse formation could take over the SA node's role as pacemaker for the heart.

Conduction System

Impulses normally are transmitted through the heart by a specialized network of fibers called the conduction system. The electrical impulse that originates in the SA node spreads across the atrial muscle and causes the atria to contract first. Within the interatrial septum near the junction of the atria and ventricles is another specialized group of cells, the atrioventricular (AV) node. When the impulse from the SA node reaches the AV node, it is momentarily slowed before passing on to the bundle of His. The bundle of His, also known as the AV bundle, is located in the upper portion of the ventricular septum. The bundle of His divides into two bundle branches that travel down the septum and give rise to a fine network of conductive tissue called the Purkinje fibers. These, in turn, stimulate all parts of the ventricle to contract. The AV node can initiate its own impulses but usually at a much slower rate (40–60 beats per minute) than the SA node.

The Cardiac Cycle

The heart has both a "resting" and "contracting" stage. In the resting or polarized state, cardiac muscle cells have a negatively charged interior and a positively charged surface. This electrical state is the result of the concentration of intracellular potassium cations and extracellular sodium cations. When cardiac muscle cells are electrically stimulated, they depolarize. Depolarization results in cell contraction. After depolarization, the cardiac muscle cell recovers. The recovery process is called repolarization.

The electrical process that causes the heart to contract is called depolarization. The electrical process that occurs after depolarization in the recovery or resting period is called repolarization. A new impulse from the SA node normally will arrive to cause depolarization when repolarization is complete.

The cardiac cycle is the combination of the contraction (depolarization) and resting or recovery (repolarization) stages. The contraction stage is also known as systole. During the resting or recovery stage, the heart fills with blood. The resting stage is also known as diastole.

The Electrocardiogram

The heart's electrical activity is transmitted outward to the body surface. By using electrodes on the surface of the skin, the heart's electrical activity can be recorded and evaluated. An ECG is a graphic recording from the body surface of the electrical currents generated by the heart. This is accomplished by amplifying 3,000 times or more the electrical current originating from the action of the heart. The ECG is a tracing representing the electrical events of the heart. The tracing can be recorded on paper or displayed on a computer monitor.

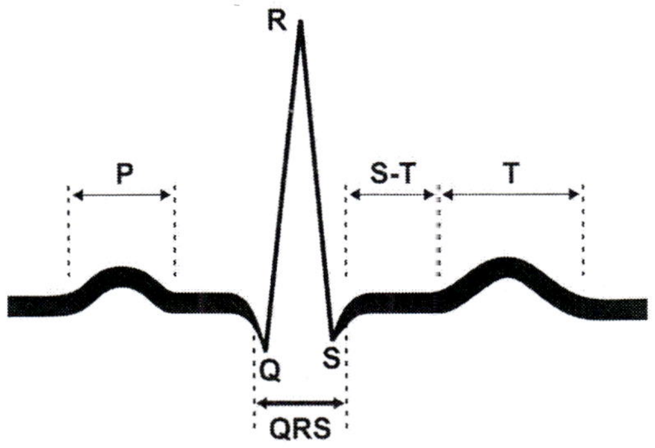

Figure 22.1 Basic electrocardiograph wave

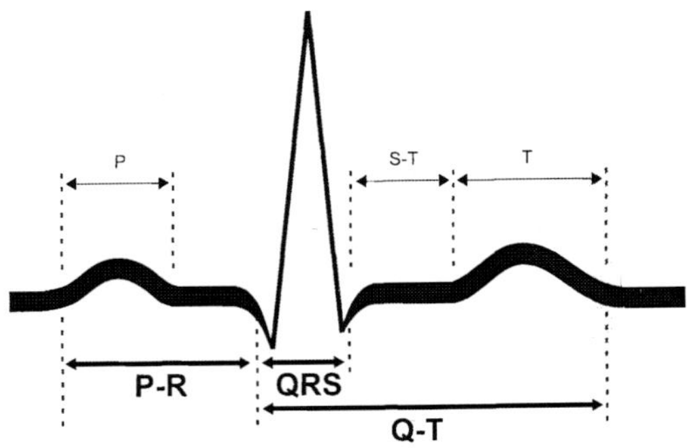

Figure 22.2 Electrocardiograph wave intervals

The ECG is used to identify cardiac rate or rhythm abnormalities and the presence of abnormal transmission of impulses through the conduction system of the heart. ECG can be performed when the heart is at rest or during exercise or pharmacologic stress. Indications for obtaining an ECG include evaluation of chest pain, cardiac rhythm disturbances, routine physical examination, and preoperative evaluation.

The Basic Tracing

Each cardiac cycle produces a complex waveform. The waveform is a series of up and down deflections from a baseline or isoelectric line. The deflections that go up are positive, and the deflections that go down are negative. Each deflection from the baseline has a letter notation: P, Q, R, S, or T. These letters were chosen arbitrarily. The basic ECG wave tracing is seen in Figure 22.1. Table 22.1 lists the ECG wave components.

The direction, size, and shape of each deflection have significance. The first deflection from baseline noted on the basic waveform is the P wave. The P wave is a positive deflection and corresponds to atrial depolarization resulting in atrial contraction. The next deflection, the Q wave, is a small negative deflection. The Q wave precedes the R wave, which is a large positive deflection. The S wave, another small negative deflection, follows the R wave. The Q, R, and S waves together are called the QRS complex and correspond to ventricular depolarization resulting in ventricular contraction. The T wave follows the QRS complex. The T wave is a positive deflection and corresponds to ventricular repolarization and relaxation. The ST segment is the time interval from the end of ventricular depolarization to the beginning of ventricular repolarization. It normally should be at baseline.

TABLE 22.1

Basic Electrocardiographic Wave Components

P wave: the part of the electrocardiograph (ECG) tracing that indicates atrial depolarization, which causes atrial contraction.

T wave: ventricular repolarization and ventricular relaxation.

QRS complex: the part of the ECG tracing that indicates ventricular depolarization, which causes ventricular contraction.

ST segment: from the end of ventricular depolarization to the beginning of ventricular repolarization.

TABLE 22.2

Standard Leads for 12-Lead Electrocardiography

Limb leads	Chest leads
I	V1
II	V2
III	V3
AVR	V4
AVL	V5
AVF	V6

It is important to be familiar with three intervals in the cardiac cycle (Fig. 22.2). The PR interval is measured from the beginning of the P wave to the beginning of the QRS complex. The PR interval is the time between atrial depolarization and the beginning of ventricular depolarization. During the PR interval, the impulse travels from the SA node to the ventricle. The QRS complex is ventricular depolarization and corresponds to the interval from the beginning of the Q wave to the S wave. If the Q wave is absent, then the QRS complex is measured from the beginning of the first positive deflection after the PR interval. The QT interval is the summation of time corresponding to ventricular depolarization and ventricular repolarization. This interval starts at the beginning of the Q wave and ends at the end of the T wave. The ST segment is an important element of the QT interval. It represents the heart's resting period between ventricular depolarization and repolarization.

The Electrocardiograph Tracing Paper

The standard ECG paper is a graph paper with divisions 1 mm apart both in width and height. The horizontal line represents time and is divided by vertical lines 1 mm apart. The heavier lines are 5 mm apart. The amount of time represented by the distance between heavy lines is 0.2 seconds. The smallest division is 0.04 seconds.

The standard paper tracing speed for the ECG machine is 25 mm per second. When the patient has a very rapid heart rate, the paper tracing speed can be increased to 50 mm per second.

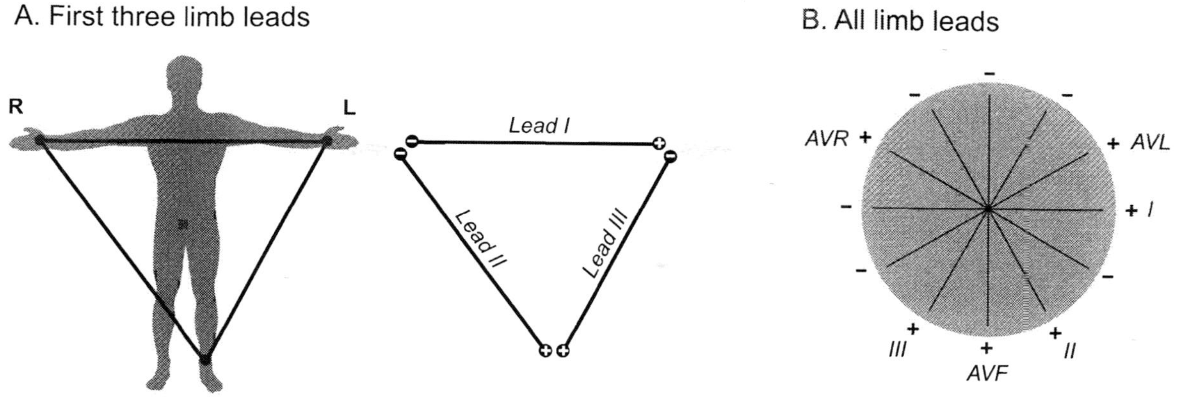

Figure 22.3 Limb leads. (A) First three limb leads. (B) All limb leads.

Standard 12-Lead Electrocardiograph Monitoring

The electrical forces generated by the heart travel in multiple directions. A complete picture of the heart's activity is obtained using a 12-lead ECG. Electrodes placed on the skin pick up electrical impulses and transmit them to the ECG recording device. As discussed in chapter 17, the skin must be properly cleansed and prepared before lead placement. The wires that connect the electrodes to the ECG machine are labeled or color-coded to ensure proper lead connection and monitoring.

The standard 12-lead ECG consists of six limb leads and six chest leads (Table 22.2). To obtain limb leads, electrodes are placed on the right arm, left arm, and left leg. Limb leads are placed at the top of the arms near the shoulders and the top of the leg. The first three limb leads are formed by a pair of electrodes; one is always positive and the other is always negative. The first three limb leads, also known as the standard limb leads, are designated as I, II, and III (Fig. 22.3A).

The next three limb leads are called AVR, AVL, and AVF. AVR records the electrical activity from midway between the left arm and left leg to the right arm. In the AVR lead, the positive electrode is on the right arm. The AVL lead represents electrical activity from midway between the right arm and left leg to the left arm. In the AVL lead, the positive electrode is on the left arm. The AVF represents electrical activity from the midpoint between the right and left arm to the left leg. In the AVF lead, the positive electrode is on the left leg (Fig. 22.3B).

The chest leads are numbered V1–V6. For each chest lead, an electrode is placed at a specific position on the chest. Lead V1 is placed just to the right of the sternum, and lead V2 is placed just to the left of the sternum. Leads V1 and V2 are the right heart leads. Leads V3 and V4 are placed next to lead V2 and are the septal leads. Leads V5 and V6 are placed laterally and over the left side of the heart. They are also known as the lateral leads. Figure 22.4 shows the placement of the chest leads. All chest leads are considered positive.

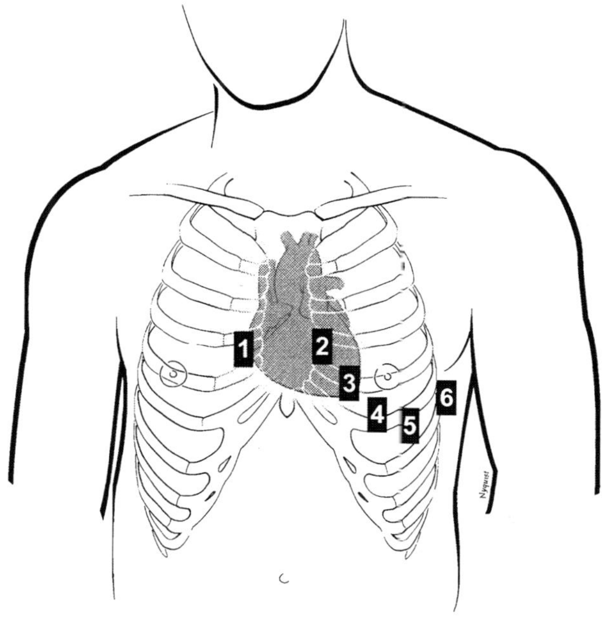

Figure 22.4 Placement of chest leads.

The ground electrode is placed on the right leg and serves as a ground.

The Electrocardiograph Tracing from Different Leads

The same cardiac activity is recorded in each lead. Each lead, however, will produce a different ECG wave or pattern based on its electrode's position on the body. Figure 22.5 is an example of a normal 12-lead ECG tracing. When electrical current flows toward a positive pole, the ECG will record an upward deflection from the baseline. When electrical current is flowing away from a positive pole, a downward deflection from the baseline will be recorded.

Cardiac Monitoring

Continuous cardiac monitoring on a display screen for identification of disturbances in cardiac rate, rhythm, or conduction can be accomplished using just three electrodes.

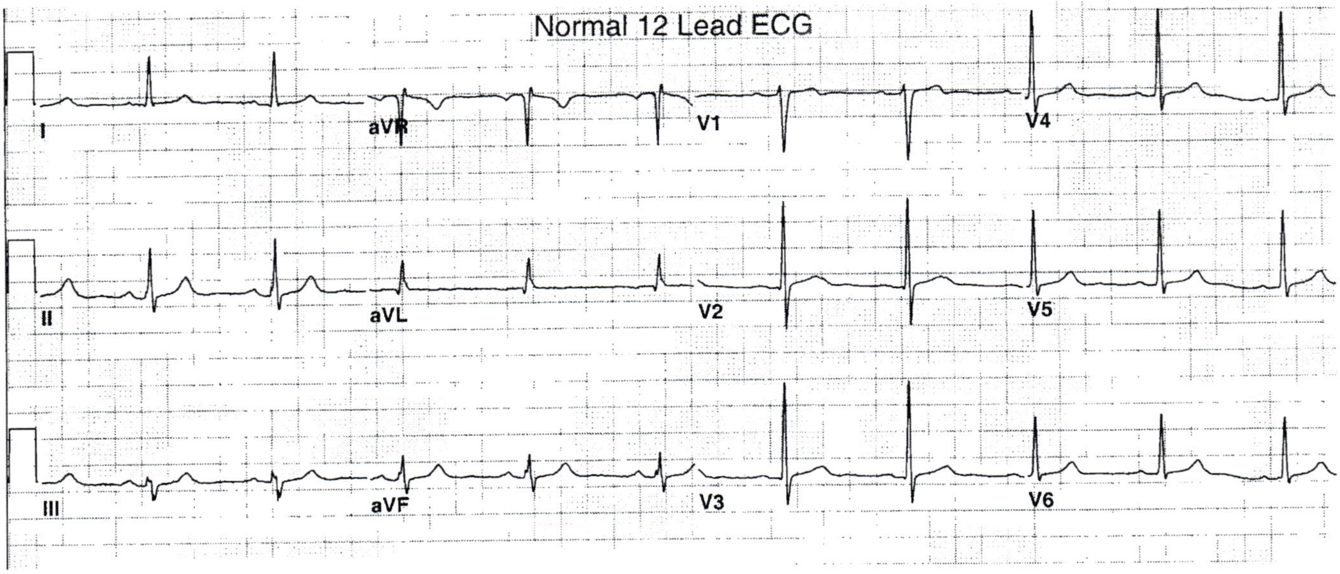

Figure 22.5 Example 12-lead electrocardiograph.

Two of three are used to detect electrical activity, and one serves as a ground. The position of the leads is chosen to provide the best ECG with a good R wave for cardiac monitoring. The conventional electrode placement for cardiac monitoring and several other variations are shown in Figure 22.6. Cardiac monitoring, although useful in certain settings, has its limitations and technical problems. It does not provide as complete a picture of the heart's electrical activity as would be seen with a 12-lead ECG.

Heart Rate and Rhythm

Heart Rate

As previously noted, the SA node is the natural pacemaker of the heart. The normal heart rate is 60–100 beats per minute. A normal heart rate and rhythm originating from the SA node are called the sinus rhythm. Sinus bradycardia occurs when the SA node paces the heart slower than 60 beats per minute. Sinus tachycardia occurs when the SA node paces the heart faster than 100 beats per minute.

If the SA node is not functioning properly, ectopic foci can assume the pacing activity. Ectopic foci for pacing are located in the atria, ventricles, and the AV junction. These foci do not normally initiate electrical activity. When ectopic foci in the atria assume heart pacing, the rate is around 60–80 beats per minute. When other ectopic foci assume pacing, the heart rate will be significantly slower than normal.

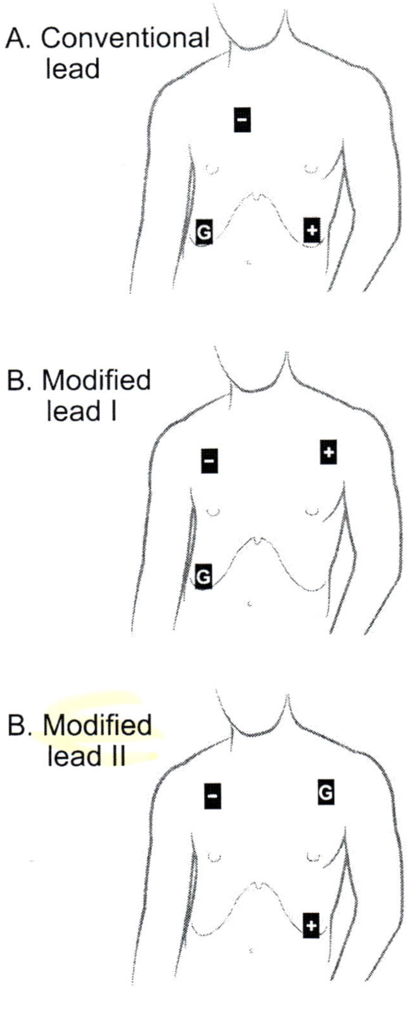

Figure 22.6 Lead positions for cardiac monitoring. (A) Conventional lead position. (B) Modified lead I. (C) Modified lead II.

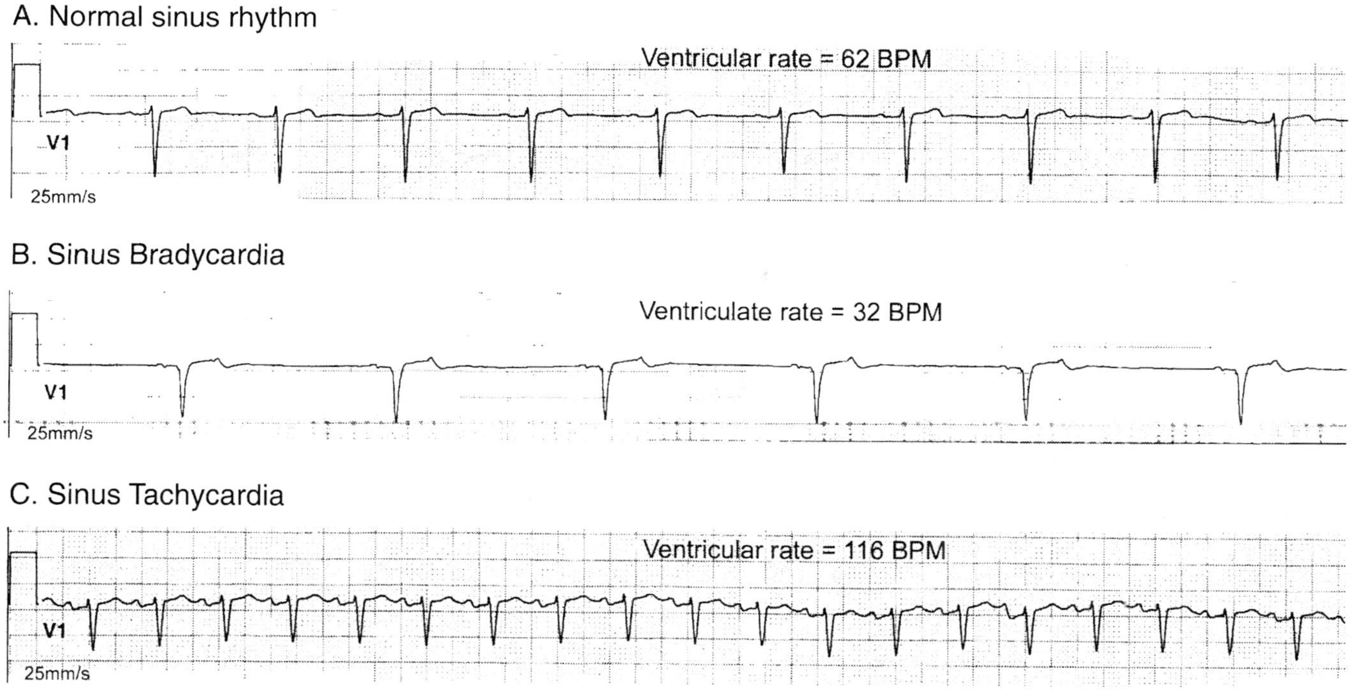

Figure 22.7 Example rhythm strips. (A) Normal sinus rhythm. (B) Sinus bradycardia. (C) Sinus tachycardia.

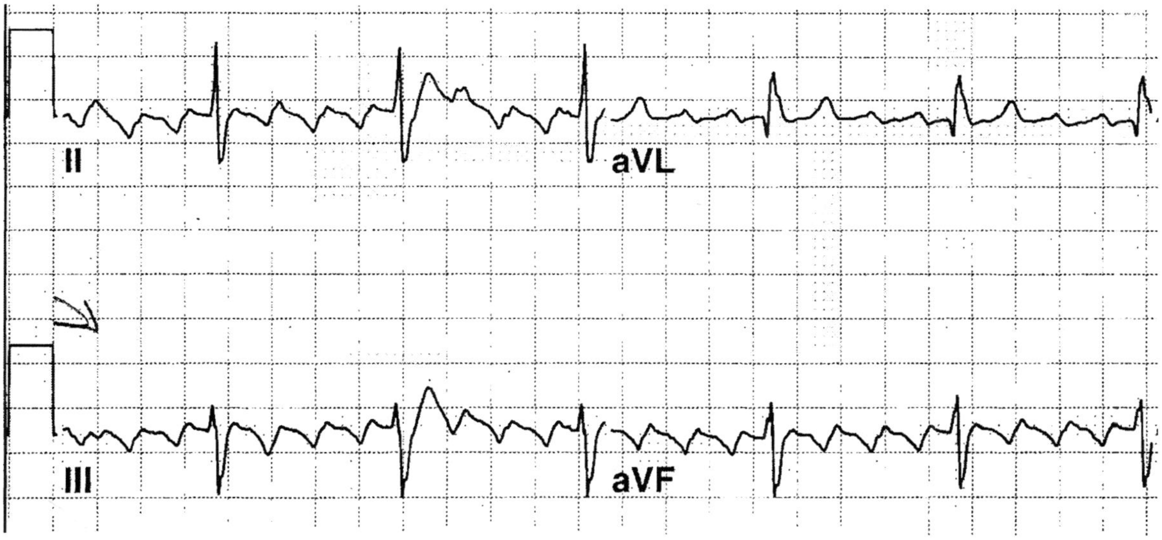

Figure 22.8 Atrial flutter.

Rate Determination

There are several ways to determine heart rate from a strip of ECG paper or multiple lead printout. If the top of the ECG paper is marked off in 3-second intervals, heart rate can be determined by counting the number of R waves in any 6-second interval and multiplying that number by 10. Another way is to measure the distance in seconds between two consecutive R waves and then divide that number into 60.

Heart Rhythm

Normal regular rhythm occurs when there is equal distance or time between similar R-R intervals. Figure 22.7 shows an example of normal sinus rhythm. Arrhythmia means irregular rhythm or R-R intervals. Irregular rhythms are usually the result of aberrations in normal pacing and/or other ectopic foci.

A number of factors may cause arrhythmia, a fast or slow heart rate, or occasional skipping of a beat. Some of

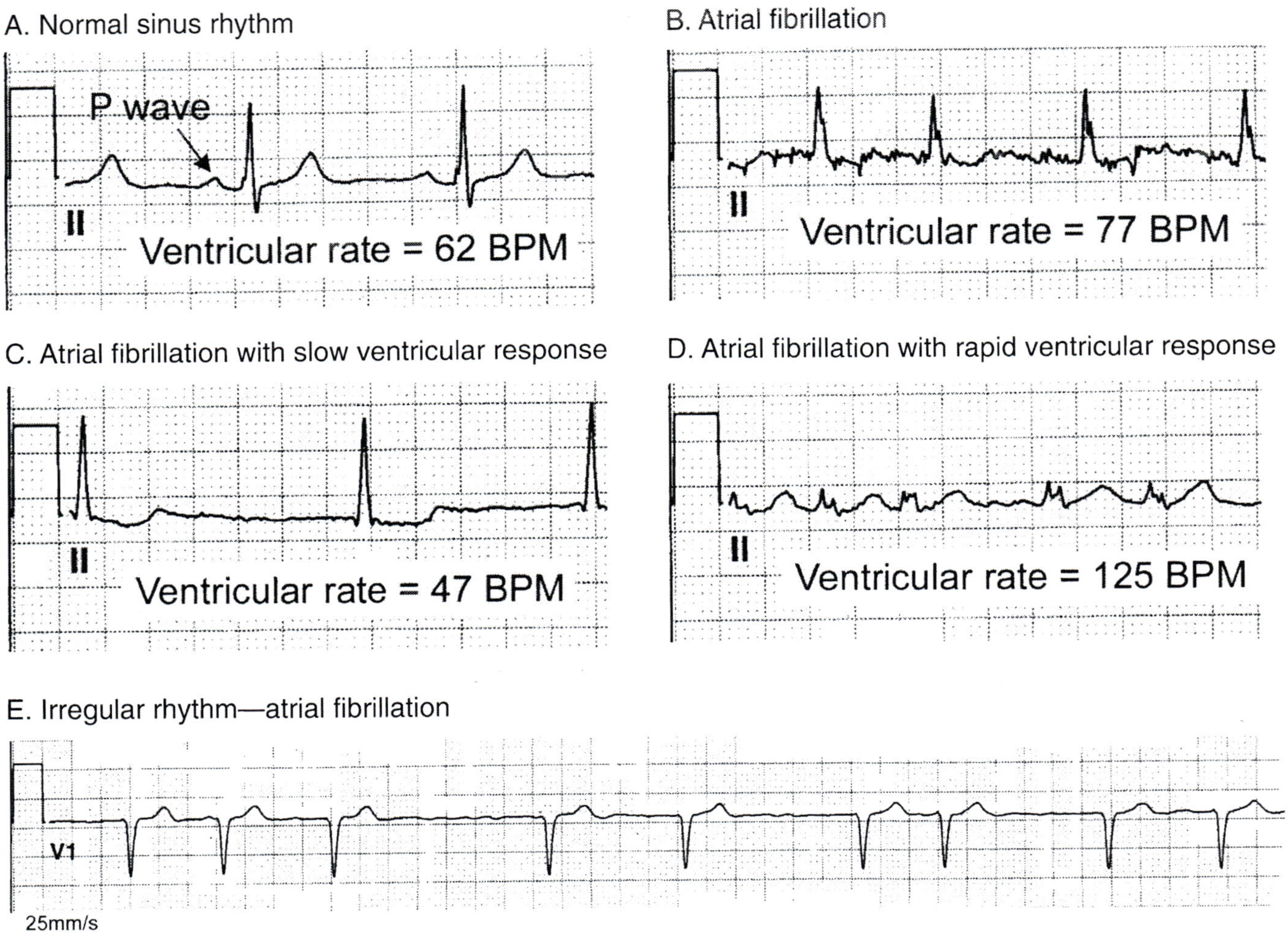

Figure 22.9 Atrial fibrillation. (A) Normal sinus rhythm. (B) Atrial fibrillation. (C) Atrial fibrillation with slow ventricular response. (D) Atrial fibrillation with rapid ventricular response. (E) Irregular rhythm: atrial fibrillation. Note the absence of a P wave in (B) and (C).

these factors include heart disease, smoking, caffeine, medications, alcohol, stress, and other factors. Not all arrhythmias are serious enough to require intervention, but some arrhythmias may require treatment or immediate attention. The type of treatment is dependent on the underlying cause. A discussion of treatment options for arrhythmias is beyond the scope of this text.

Common Abnormal Rate and Rhythm Patterns

The heart rates associated with tachycardia, flutter, and fibrillation are listed in Table 22.3. Atrial flutter is noted on the ECG by its characteristic sawtooth appearance. For an example of atrial flutter, see Figure 22.8.

Atrial fibrillation is caused by uncontrolled, continuous firing of multiple foci in the atria. This results in an ineffective quivering of the cardiac muscle. It is characterized by irregular ventricular rhythm and absence of a P wave. When the ventricular response is rapid with prematurely conducted ventricular complexes, it may be difficult to get an adequate pacing trigger for gated cardiac studies. For examples of atrial fibrillation, see Figure 22.9

Premature ventricular contractions (PVCs) originate from an ectopic focus in the ventricle and produce a wide ventricular QRS complex. PVCs can be unifocal or multifocal (such as a couplet, triplet, or run of PVCs) and can occur as bigeminy or trigeminy. A repeating pattern of PVCs

TABLE 22.3
Abnormal Heart Rates

Normal sinus rhythm	60–100 beats/min
Tachycardia	>100 beats/min
Flutter	250–350 beats/min
Fibrillation	>350 beats/min

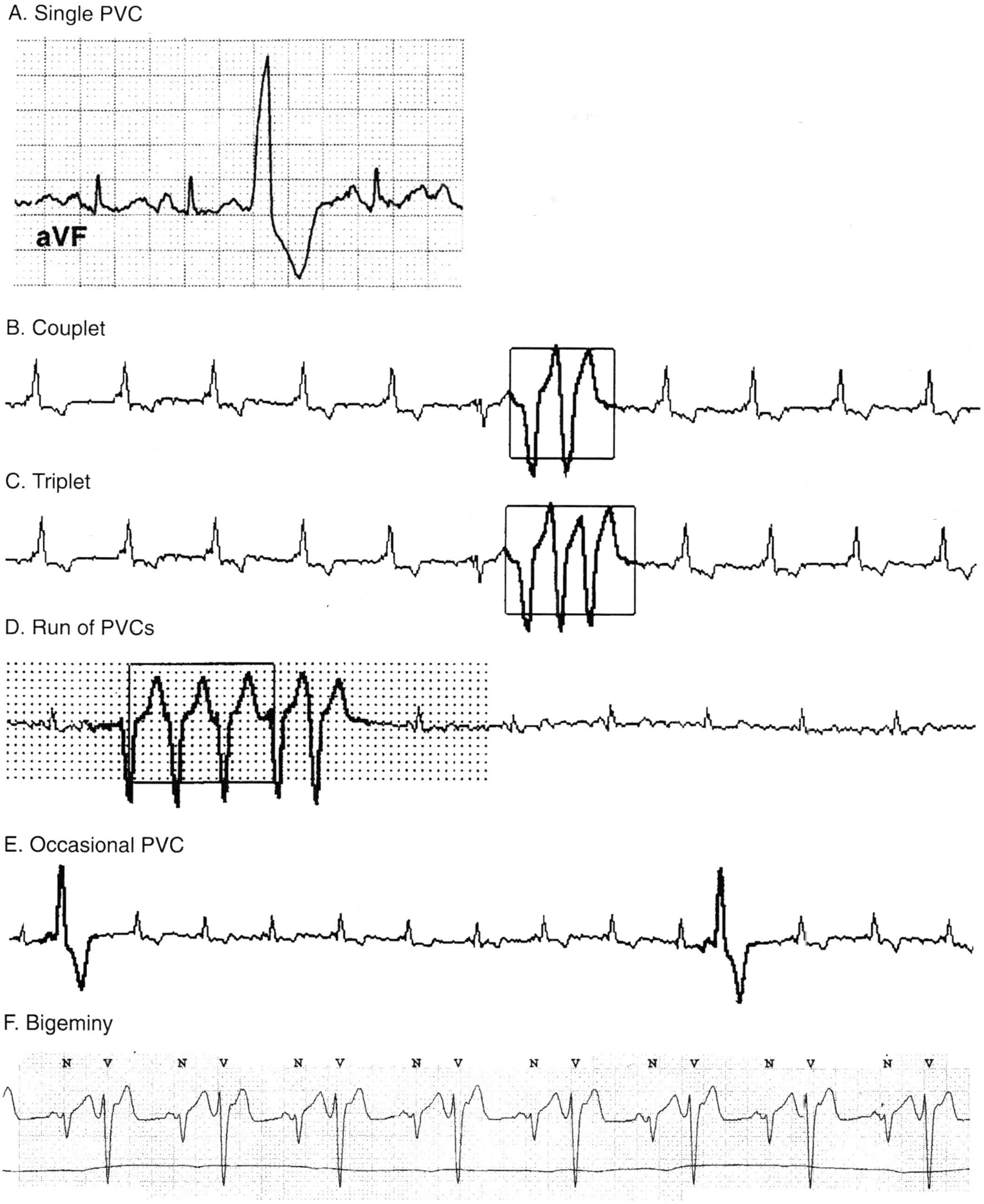

Figure 22.10 Premature ventricular complexes (PVCs). (A) Single PVC. (B) Couplet. (C) Triplet. (D) Run of PVCs; also called a short run of ventricular tachycardia. (E) Occasional PVC. (F) Bigeminy. N = normal beat; V = PVC.

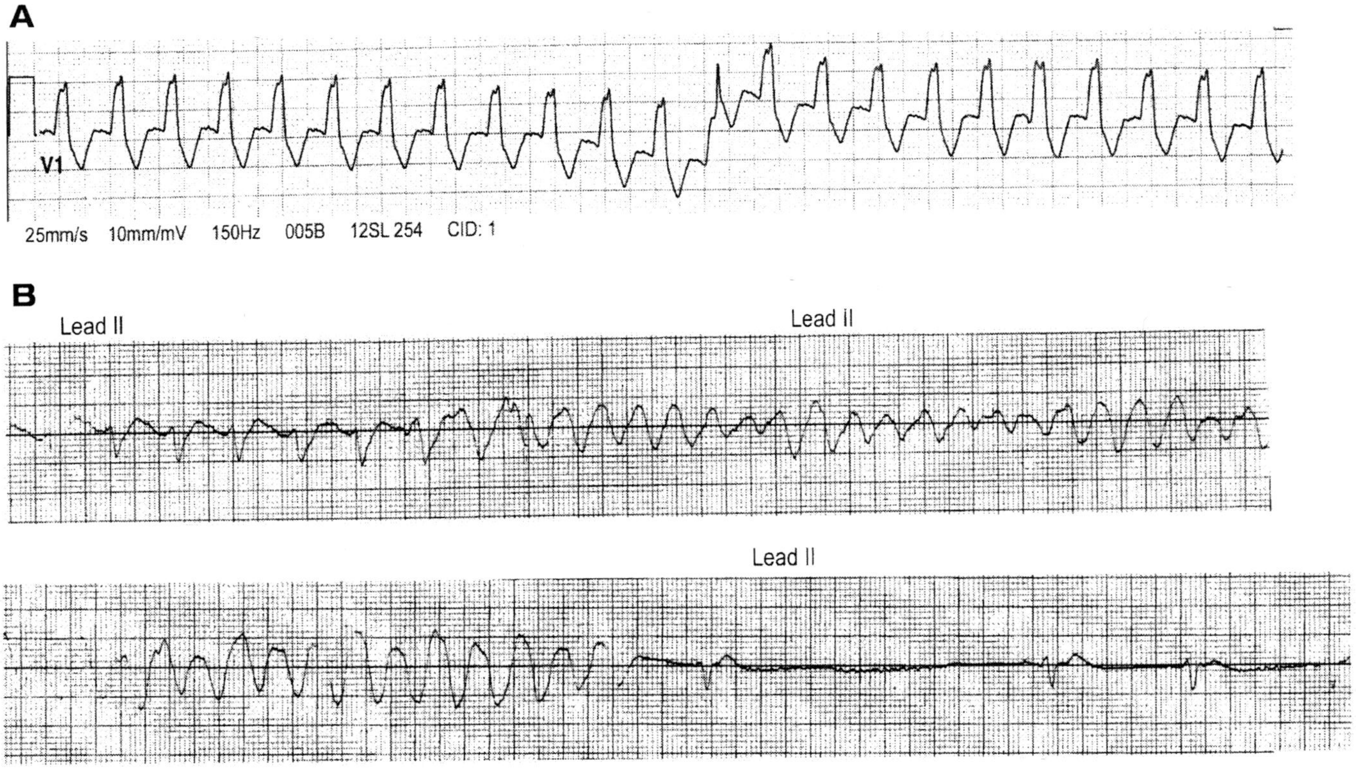

Figure 22.11 Ventricular tachycardia and fibrillation. (A) An example of ventricular tachycardia shows QRS complexes in rapid sequence. (B) Both strips are examples of ventricular fibrillation. In the second strip, the ventricular fibrillation is followed by asystole.

followed by a normal beat is called ventricular bigeminy. Trigeminy is a pattern of PVCs followed by two normal beats. Examples of PVCs are shown in Figure 22.10

The appearance of three or more rapid consecutive PVCs is known as ventricular tachycardia. If not controlled, ventricular tachycardia can lead to ventricular flutter or fibrillation, which are life-threatening situations. In ventricular flutter/fibrillation, the ECG tracing becomes wavy and irregular, with no discernible QRS complex or P wave (Fig. 22.11)

Heart Blocks/Conduction Disruption

Electrical conduction disorders from the SA node, AV node, and Purkinje fibers result in heart blocks. The conduction interference may be partial or complete. Heart blocks are classified by the extent of the conduction abnormality. In a first-degree heart block, the electrical impulse is conducted more slowly than normal through the affected zone. In a second-degree heart block, the electrical impulse may or may not be conducted. In a third-degree heart block, the electrical impulse is totally blocked. This classification is used with all types of heart blocks.

Examples of heart blocks include sinus block, atrioventricular block, and bundle branch block (BBB). A complete discussion of heart blocks is beyond the scope of this text. The appearance of the P wave and the QRS complex will vary with different types of heart blocks.

In a sinus block, the SA node may temporarily fail to pace. This type of block is recognized by absence of the P wave and QRS complex.

An AV block delays the stimulation of the ventricles. In a first-degree AV block, the PR interval is prolonged beyond 0.02 seconds because of the delay in conduction of the impulse through the AV node. In second-degree AV block, the PR interval becomes gradually longer and the QRS complex fails to occur, resulting in an occasional "dropped" or absent QRS complex. Third-degree AV blocks occur when none of the atrial impulses is conducted to the ventricles and the ventricles pace independently.

The most frequent location of blocks originating in the ventricles is in the trunk of the right or left bundle branches. BBBs are caused by a block of depolarization in the right or left bundle branches. In a BBB, the peak of the QRS complex is notched. The QRS complex must be wide enough to make the diagnosis of BBB. An example of the notch is seen in Figure 22.12. The R' represents the delay in polarization caused by the BBB. Figure 22.13 shows an example of the QRS configuration in right and left BBBs.

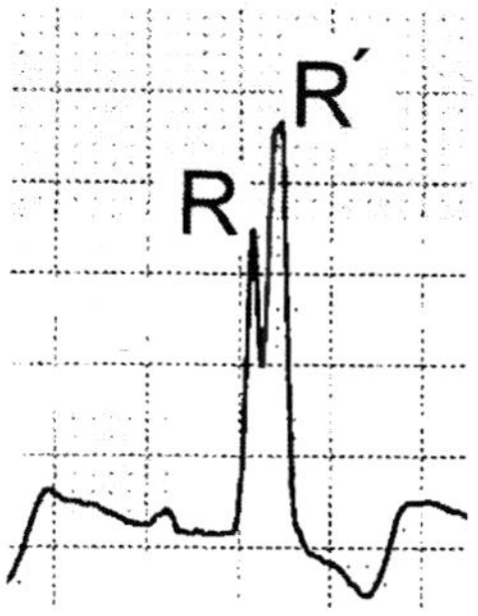

Figure 22.12 Notched appearance of QRS in bundle branch block.

Ischemia/Infarction

ST segment depression may be seen in subendocardial infarction, in patients on digitalis, and in transient exercise-induced ischemia. Figure 22.14 is an example of ST segment depression.

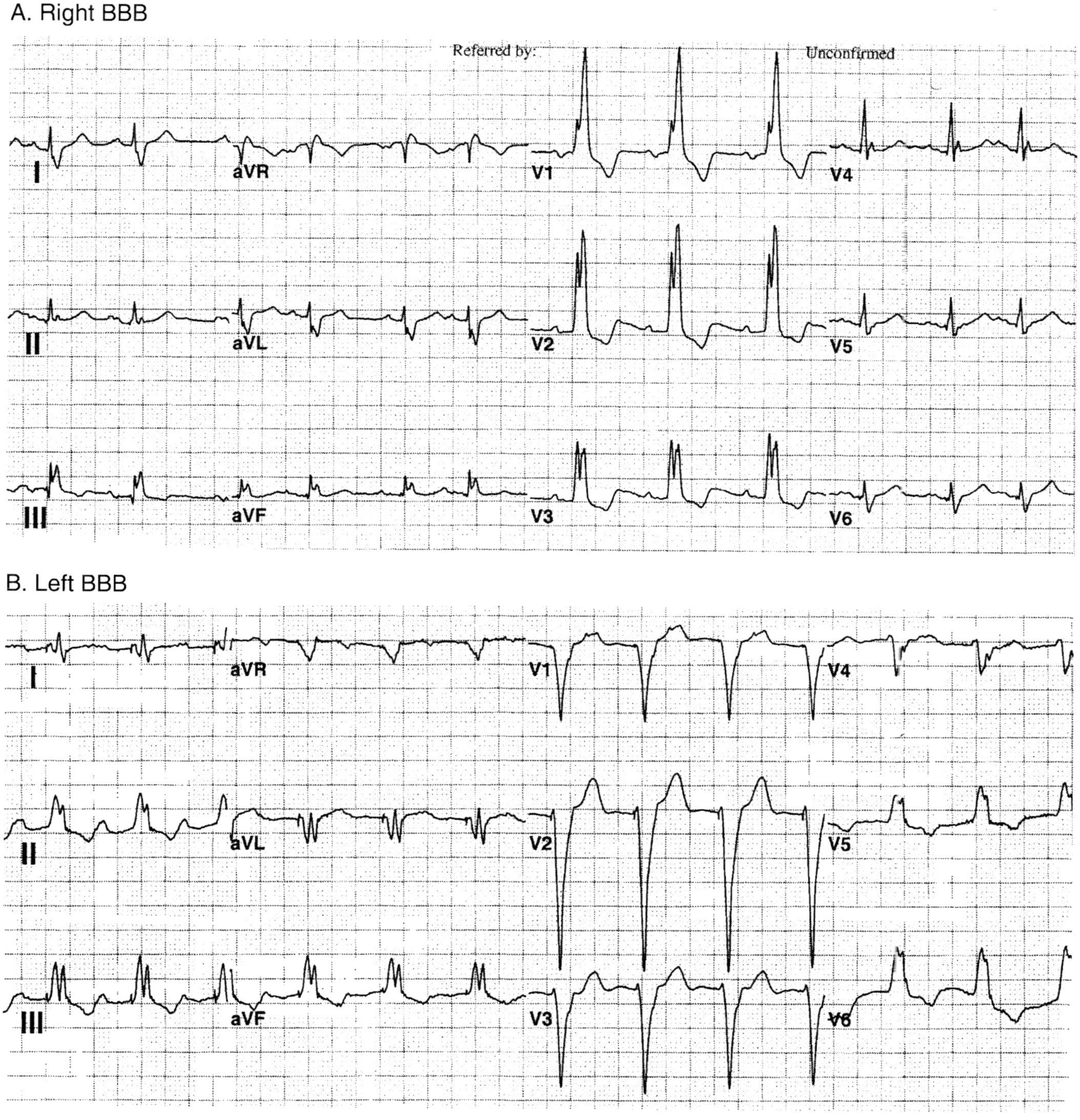

Figure 22.13 Examples of bundle branch block (BBB). (A) Right BBB. (B) Left BBB.

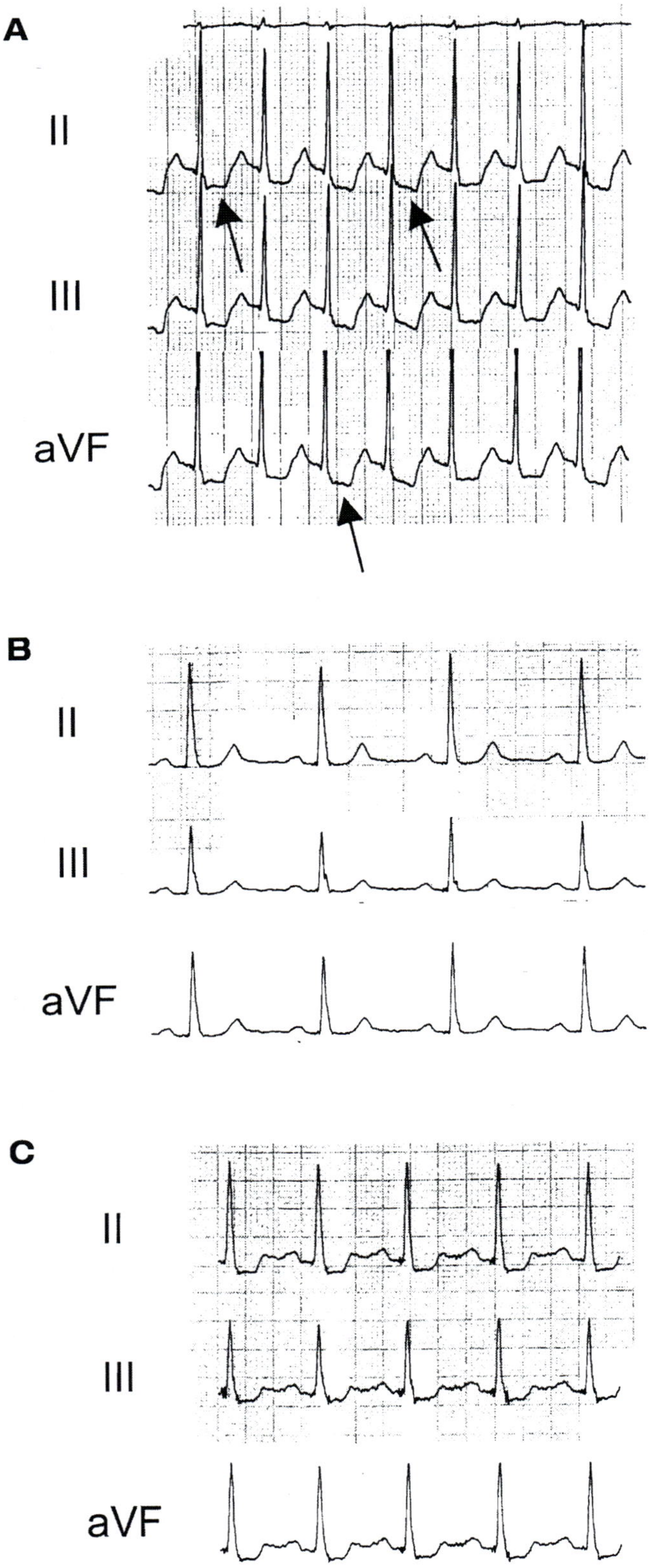

Figure 22.14 ST segment depression. (A) The arrow indicates the ST segment depression. (B) Electrocardiograph (ECG) from resting patient. (C) ECG at peak exercise in the same patient. ST segment depression is evidence of stress-induced ischemia.

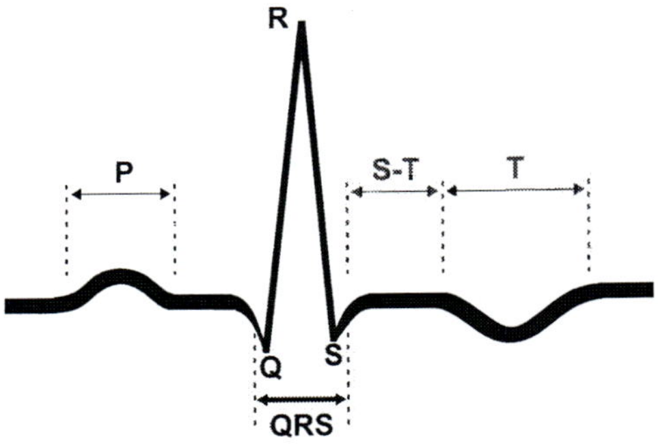

Figure 22.15 T wave inversion

T wave inversion represents ischemia. Its appearance may be anything from moderately flattened to significantly inverted. Figure 22.15 is a diagram of T wave inversion.

ST segment elevation represents acute ischemia resulting in injury. The degree of the ST segment elevation denotes the severity of the ischemic injury. Figure 22.16 is an example of ST segment elevation.

The presence of a Q wave denotes myocardial infarction. The Q wave normally is not notable. All 12 leads must be checked for the presence of a Q wave. The anatomical location of an infarct is determined by the presence of a Q wave in a given lead. An example of infarction is seen in Figure 22.17

Myocardial Segments Related to Electrocardiographic Changes

The myocardial segment(s) corresponding to ECG abnormalities is determined by the leads in which the abnormality is seen. Table 22.4 lists the leads corresponding to the different myocardial segments. Sometimes an infarct extends into an adjacent territory. For example, when an inferior wall infarct extends into the lateral wall, it is known as an inferiolateral infarct. Other common territorial combinations include anteroseptal, anterolateral, anterobasal, and posterolateral walls. A combination of ECG leads of the respective myocardial walls will depict the corresponding abnormalities.

TABLE 22.4

Electrocardiograph Leads Corresponding to Myocardial Segments

Wall or segment	Leads
Anterior	I, II
Lateral	V, VI
Septal	III, IV
Inferior	II, III, AVF
Inferolateral	II, III, AVF, V, VI

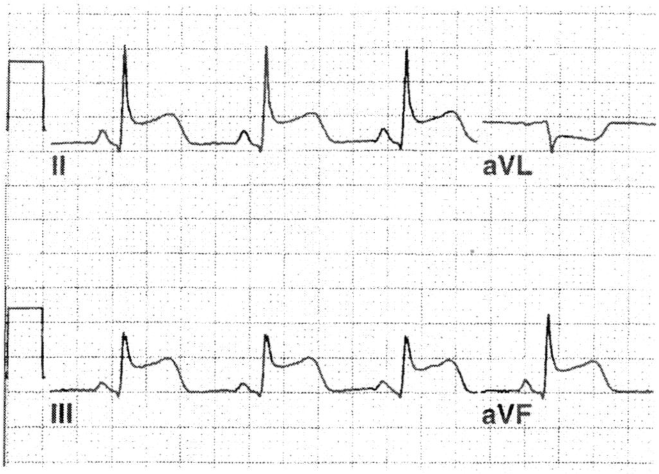

Figure 22.16 ST segment elevation. ST segment elevation denoting inferior injury or acute infarction.

Artificial Pacemakers

An artificial pacemaker is an electronic device implanted in the chest and can stimulate the heart to beat when the electrical conduction system of the heart is unable to function properly. An artificial pacemaker can be used for permanent or temporary pacing. A pacemaker can sense the electrical activity of the heart and deliver electrical impulses to the atria and/or ventricles as needed. A pacemaker can be used for atrial, ventricular, or dual-chamber pacing. The pacemaker consists of an electronic pulse generator, a battery, and a wire lead for sensing and conduction of electrical impulses. The artificially generated electrical impulse from a pacemaker causes a narrow line, known as the pacemaker spike, in the ECG tracing. Examples of pacemaker spikes are seen in Figure 22.18.

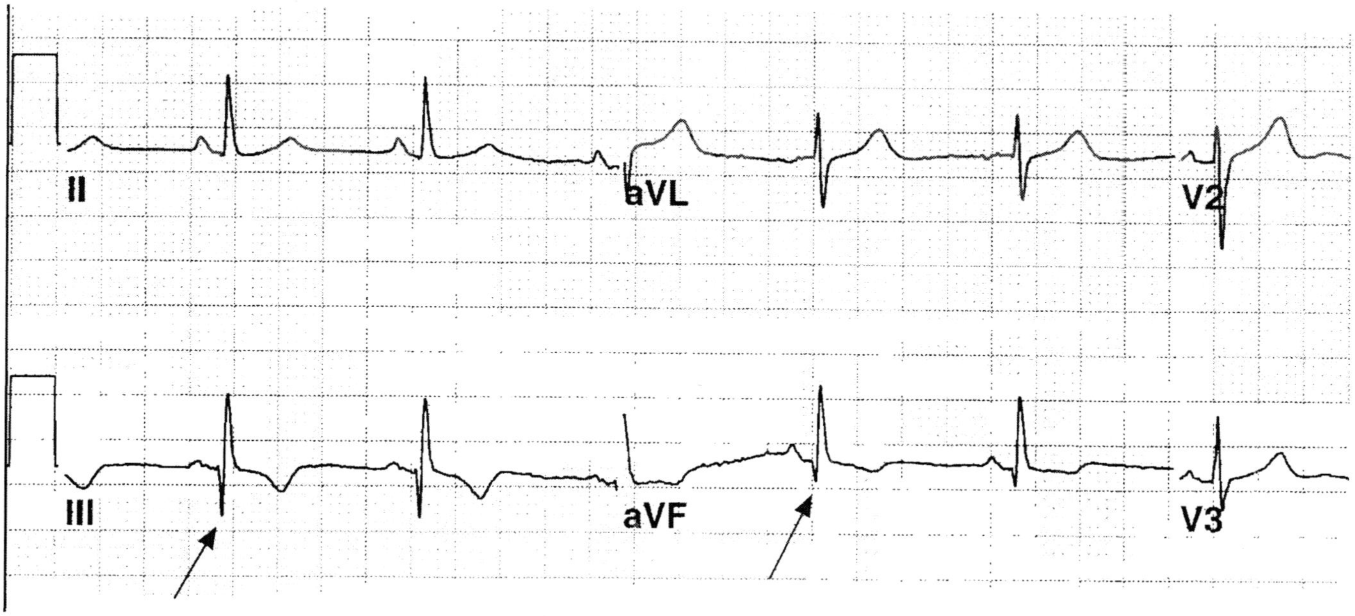

Figure 22.17 Q wave and infarction. Possible inferior infarct, age undetermined. Arrows point to Q wave. Also notice T wave inversion.

A. Atrial pacing

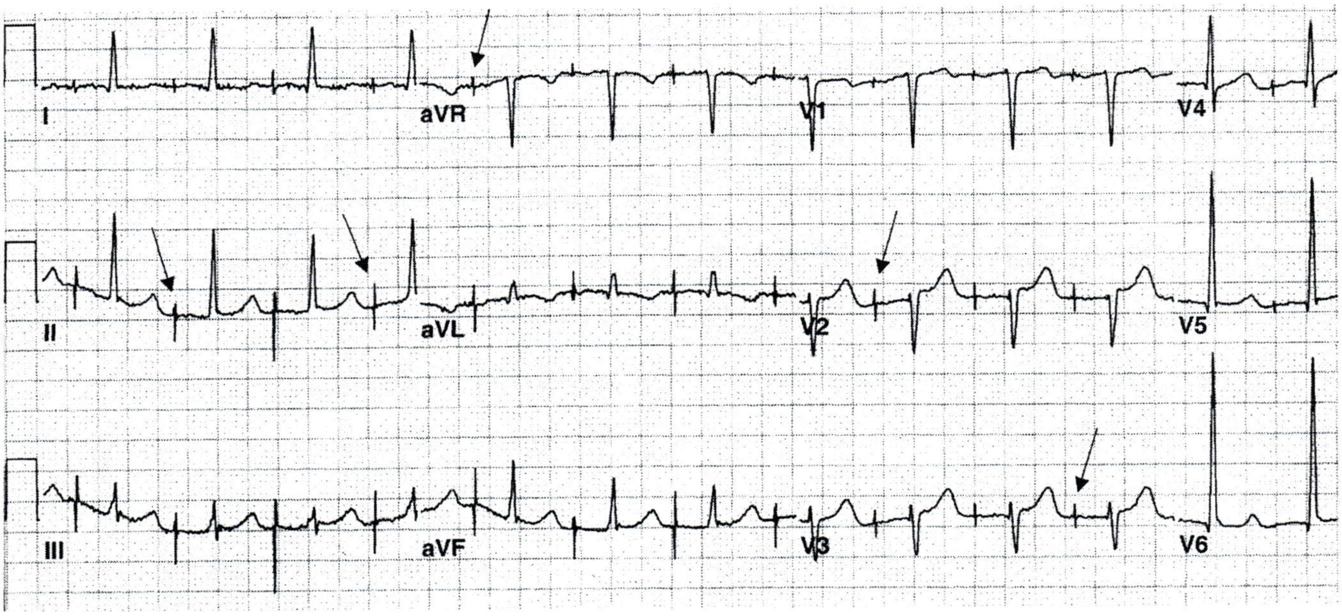

B. Ventricular pacing

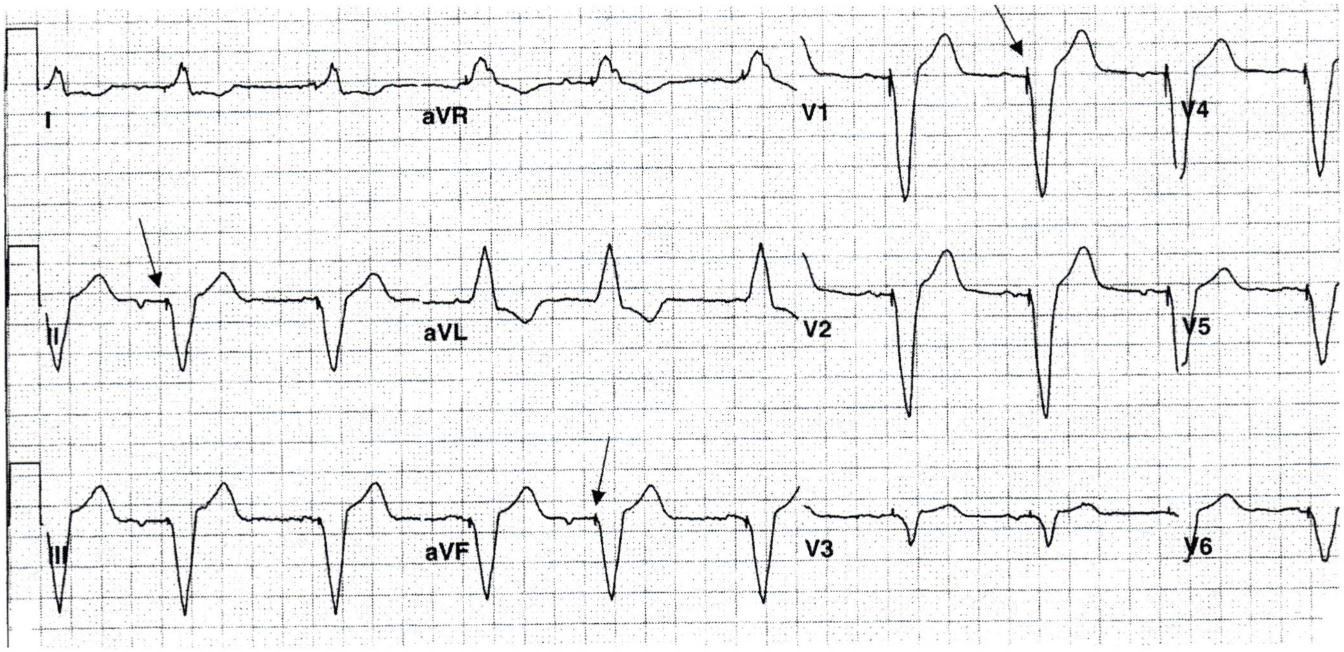

Figure 22.18 **Pacemaker spikes. (A) Atrial pacing. (B) Ventricular pacing.**

C. Dual pacing

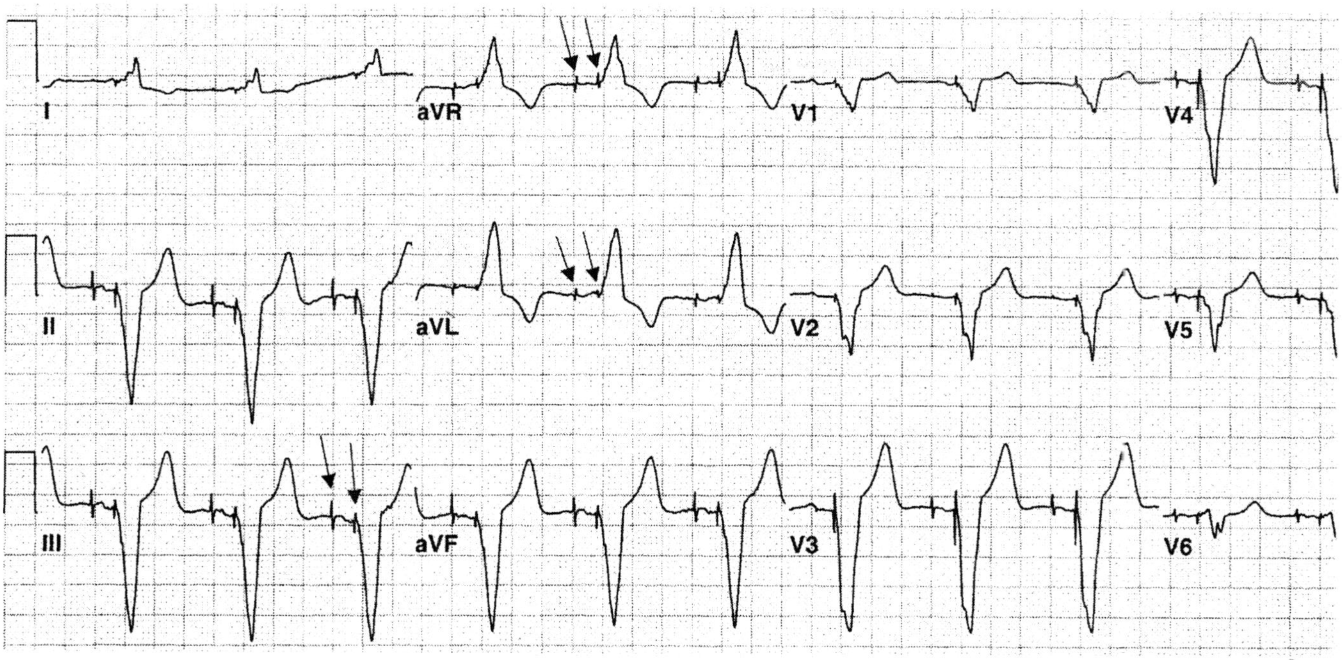

D. Large pacemaker spike

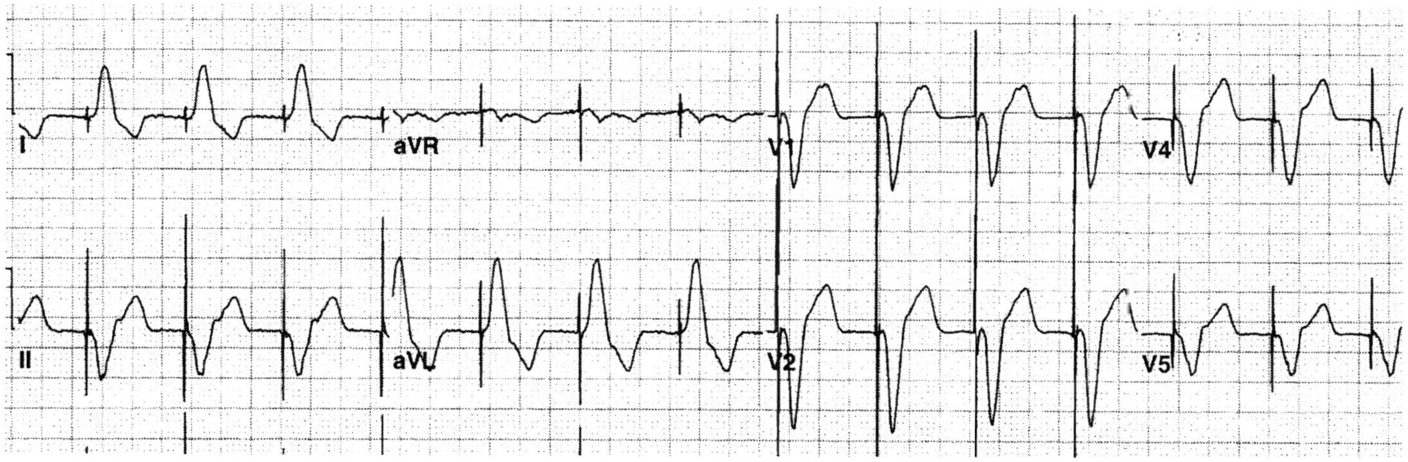

Figure 22.18 (C) Dual pacing. (D) Large pacemaker spike. Arrows point out some of the pacemaker spikes seen in different leads. A large pacemaker spike as seen in (D) can interfere with gated acquisition.

References

1. Huszar RJ. *Basic Dysrhythmias. Interpretation and Management.* 3rd ed. St. Louis, MO: Mosby; 2001.
2. Bayes de Luna A. *Clinical Electrocardiography: A Textbook.* 2nd ed. Armonk, NY: Futura Publishing Co.; 1998.
3. Atwood S, Stanton C, Storey J. *Introduction to Basic Cardiac Dysrhythmias.* 2nd ed. St. Louis, MO: Mosby-Year Book, Inc.; 1996.
4. Dubin D. *Rapid Interpretation of EKGs.* 4th ed. Tampa, FL: Cover Publishing Co; 1989.
5. Surawicz B, Knilans T, Chou T-C. *Chou's Electrocardiography in Clinical Practice. Adult and Pediatric.* 5th ed. Philadelphia, PA: WB Saunders Co; 2001.
6. Vinsant MO, Spense MI. *Common Sense Approach to Coronary Care. A Program.* 3rd ed. St Louis, MO: Mosby; 1981.